AF816900

SOTHIS PRESS

Sothis Press aims to completely reimagine the public domain book with an emphasis on universal access to captivatingly designed print editions. We believe that technology will enable every single book ever published to be perpetually accessible in beautiful print form. Sothis is beauty in ubiquity.

www.sothis.press
1 Waterhouse Square, London, ECIN 2ST

The Teacher's Handbook of Slöjd

As Practised and Taught at Naäs, Containing Explanations and Details of Each Exercise

Otto Salomon

SOTHIS PRESS

ISBN 978-1-0365-1177-7

The Teacher's Handbook of Slöjd
is in the public domain.

THE TEACHER'S

HAND-BOOK OF SLÖJD

AS PRACTISED AND TAUGHT AT NÄÄS

CONTAINING EXPLANATIONS AND DETAILS OF EACH EXERCISE.

By **OTTO SALOMON,**

Director of the Nääs Seminarium.

ASSISTED BY CARL NORDENDAHL AND ALFRED JOHANSSON.

TRANSLATED AND ADAPTED FOR ENGLISH TEACHERS

BY MARY R. WALKER,
St. George's Training College, Edinburgh,

AND WILLIAM NELSON,
Of the Manchester Schools for the Deaf and Dumb.

WITH OVER 130 ILLUSTRATIONS AND PLATES.

SILVER, BURDETT & CO.,

PUBLISHERS,

BOSTON, NEW YORK, CHICAGO.

1891.

PREFACE TO THE SWEDISH EDITION.

A DESIRE has for some time been expressed in various quarters for a Hand-Book of Slöjd, written from the *educational* point of view. There have been many indications, especially in connection with Slöjd carpentry, that teachers are not well enough acquainted with the tools employed to select and manage them properly; and a degree of uncertainty seems to prevail regarding the right method of executing the exercises. Now, it is true that no one can acquire this knowledge from *books;* the way to acquire it is by practical, personal experience. Yet, to retain this experience, and apply it, is partially a matter of memory, and, therefore, systematically arranged directions are capable of rendering aid which is not to be despised. A hand-book like the present does not, and could not, supersede personal experience at the bench, or render a course of instruction unnecessary. Its sole object is to supplement and complete the notes which every conscientious student takes during such a course. Its aim is, therefore, chiefly to strengthen and confirm knowledge already acquired; but, though it is thus limited in scope, and, on this account, perhaps to be regarded as in some respects incomplete, the writers venture to express the hope that it will be welcomed by many teachers.

Books are, perhaps, more frequently published before their time than after it; and although there have been numerous opportunities for observation in the province of Educational Slöjd during the last eighteen years (the Slöjd Institution at Nääs having begun operations in 1872), the writers are nevertheless uncertain whether the time has really yet come for the publication of definite directions; or, at least, whether their knowledge of the subject is yet complete enough to justify their appearance in print. But, if they have been premature, the sole reason is to be found in their desire to satisfy a want, which becomes every year more pressing.

The views expressed in the book are, for obvious reasons, in full accordance with the system of instruction followed at Nääs. They are the outcome of careful observations, and of experiments tested by practice. Yet, even if these views should be confirmed by many teachers, the writers, knowing that opinions are divided in the matter of instruction in Slöjd, as in most other questions, are fully prepared for adverse criticism. Whether this criticism be justified or not, of one thing they are certain, and that is, that in all honesty of purpose and strength of conviction they have striven to fulfil a far from easy task. They trust that others with greater ability will succeed them and do it better. So little attention has hitherto been paid to the subject in question that it has been necessary to generalise and draw conclusions almost exclusively from personal experience. But their motto has been—"Prove all things, hold fast to that which is good"; and much that in the beginning and in the light of comparatively limited experience met with their approval, has, on closer examination, been rejected or modified.

But, though this hand-book is necessarily the outcome chiefly of personal observation and experience, the writers have to some extent been able to avail themselves of the knowledge of others, and to refer to competent authorities. This applies especially to Chapter II., for the contents of which frequent reference has been made to the writings of Karmarsch, Thelaus, and others. The Plates at the end, and most of the Illustrations in the body of the book, are executed from original drawings made for the purpose.

In order to keep within due limits, much has been omitted which, perhaps, ought to have been included. Whether or not, on the other hand, some things have been included which ought to have been omitted, must in the meantime be left an open question.

The parts taken by the respective authors are as follows:—Chapter I. has been written by Otto Salomon; Chapters II., III., and IV., by Carl Nordendahl, who also undertook all arrangements connected with the illustrations; and Chapter V., by Alfred Johansson. Looked at as a whole, however, this little book is the product of united labour, and it contains nothing which is not the result of diligent interchange of thought.

TRANSLATORS' PREFACE.

THIS Hand-book was written originally for Swedish people, and in accordance with the conditions which prevail in Swedish schools; but the presence of a large body of English teachers at the Autumn Slöjd Course at Nääs has testified for the last four years to the interest taken in the subject by English people, and the latest modifications of the English and Scotch Codes as regards manual training, point to the introduction at no distant date of systematic instruction in some branch of manual work in our state-aided schools. It has therefore seemed desirable that this Hand-book of Wood Slöjd should be translated for English readers with any modifications necessary to make it suitable for English teachers and students. These modifications consist partly of the omission of matter bearing on conditions peculiar to Sweden, and partly of the addition to the text of certain paragraphs, which seemed necessary from an English point of view. Nothing has been taken away or added without careful consultation with Herr Salomon, and without his approval. At the same time, as any additions to the original text have been made at the suggestion of the translators, and as they are responsible for them, these paragraphs have been enclosed in brackets as translators' notes. The whole translation has been revised under the supervision of Herr Salomon and other competent judges at Nääs, and the translators therefore trust that the work they have undertaken is a faithful representation of the views held and acted on at the headquarters of Educational Slöjd.

In giving this book to English readers, they feel, however, that one or two points of detail call for special explanation, particularly as these touch on the fundamental principles of educational Slöjd, and as any misunderstanding as to details might lead to a more serious misunderstanding as to principles. One of these details is *the use of the knife in educational Slöjd.* In the following pages the use of the knife is often recommended where the English carpenter would use the chisel, or some other special tool. The defence of the knife in such cases is to be found in the fact that, while it is the most familiar and the simplest tool which can be put into the hands of the pupil, it is full of potentialities in the hands of the intelligent worker, who can perform with it many exercises which the tradesman executes in a more mechanical way with some other tool.

Again, directions are given which differ in other respects from those which the carpenter would give. The work of the slöjder is often done not only with different tools, but in a different *order* from that of the artisan. This inversion of order is a natural consequence of the principle that each article shall be executed entirely by the individual worker. Division of labour, though necessary from the tradesman's point of view, is not permitted in Slöjd, deadening, as it does, individuality, and reducing to a minimum the calls made on the intelligence.

These and other deviations from the methods of the carpenter are made not in ignorance, but of set purpose, and have their grounds in the comprehensive principle that all method in Slöjd must aim in the first place at the physical and mental development of the pupil, and only at the production of articles in so far as this subserves the primary aim.

In close connection with this stands the question of the place occupied in the system by the articles produced, *i.e.*, by the *models*. Clear as this question appears in the light of the fundamental principles on which educational Slöjd is based, the idea still seems to prevail to some extent that, if the principles are accepted, the Nääs models must also be accepted unconditionally, and that the two stand and fall together. So far is this from being the case that, at the present time, one series of Nääs models is gradually becoming English in its character, and only waits further suggestions from English teachers to become entirely so. The sole reason that it still contains models which do not entirely fulfil the condition of being familiar and useful in the homes of English children, is that English people have hitherto been unable to suggest satisfactory substitutes. The models are merely the *expression* of the system, and to carry out that system thoroughly they must be national in their character, and ought, therefore, to vary in their nature with the countries into which Slöjd is introduced as a subject of instruction.

The translators are at present engaged on an English edition of Herr Johansson's Manual of Directions for making the models mentioned in the preceding paragraph. This Manual, which will be ready for issue shortly, will complete the Handbook on the purely *practical* side. As the principles on which Slöjd rests as an *educational factor* are necessarily very briefly dealt with in the Handbook, the translators are glad to learn that "The Theory of Slöjd," the only authorised English edition of Herr Salomon's Lectures, edited by an Inspector of Schools, will shortly appear, and will form a companion volume to this Handbook.

As this translation, like the original, is the work of more than one writer, it remains to add that the book has been translated into English by Mary R. Walker, with the assistance of William Nelson on all points relating to technical knowledge and technical terminology.

Table of Contents.

CHAPTER I.

Introductory Remarks.

CHAPTER II.

Wood or Timber.

CHAPTER III.

Tools.

CHAPTER IV.

Jointing.

CHAPTER V.

FIRST CHAPTER.

Introductory Remarks.

Educational Slöjd.

By educational slöjd is meant the application of *slöjd* to educational purposes. Slöjd is not to be confounded with the work of the artisan—a mistake which may easily happen if the distinction is not sufficiently strongly emphasized. Speaking generally, the 'slöjder' does not practise his art as a trade, but merely as a change from some other employment; and in the nature of the articles produced, in the tools used in their production, in the manner of executing the work, etc., slöjd and the work of the artisan differ very decidedly the one from the other. Slöjd is much better adapted to be a means of education, because purely economical considerations do not come forward so prominently as must be the case with work undertaken as a means of livelihood.

Educational and practical slöjd.

Educational slöjd differs from so-called *practical slöjd*, inasmuch as in the latter, importance is attached to the *work;* in the former, on the contrary, to the *worker*. It must, however, be strongly emphasized that the two terms, *educational* and *practical,* ought in no way to be considered antagonistic to each other, as frequently happens in popular language; for, from the strictly educational point of view, whatever is educationally right must also be practical, and *vice versa.* When the educational and the practical come into conflict, the cause is always to be found in the pressure of adventitious circumstances, *e.g.*, the number of pupils, the nature of the premises, and, above all, pecuniary resources, etc. To make educational theory and practice coincide is an ideal towards which every teacher must strive. One man, perhaps, may be able to come nearer to this common ideal than another, but everyone, as he runs his course, must have this goal clearly in view, and in every unavoidable compromise he must endeavour to make what *ought* to be done and what *can* be done come as close together as possible.

The aim of educational slöjd.

What, then, is the aim of educational slöjd? To utilise, as is suggested above, the educative force which lies in rightly directed bodily labour, as a means of developing in the pupils physical and mental powers which will be a sure and evident gain to them for life. Views may differ as to what is to be understood by a "cultured" or an "educated" man, but however far apart in other respects these views may lie, they all have at least one thing in common, *i.e.*, that this much disputed culture always appears in its possessors in the form of certain faculties, and that therefore the development of faculty, so far as this can be directed for good, must enter into all educational efforts. This being the case, the influence of slöjd is cultivating and educative, just in the same degree as by its means certain faculties of true value for life reach a development which could not be attained otherwise, or, at least, not in the same degree. Educational slöjd, accordingly, seeks to work on lines which shall insure, during and by means of the exercise it affords, the development of the pupil in certain definite directions. These are of various kinds. As the more important, it is usual to bring forward: *pleasure in bodily labour, and respect for it, habits of independence, order, accuracy, attention and industry, increase of physical strength, development of the power of observation in the eye, and of execution in the hand.* Educational slöjd has also in view *the development of mental power*, or, in other words, is *disciplinary* in its aim.

The Teacher of Educational Slöjd.

The qualifications required in the teacher.

That no one can teach what he does not know himself is a proposition the validity of which cannot be called in question.

It is equally incontestable that it is by no means sufficient to be in possession of a certain amount of knowledge and dexterity in order to follow with success the important and responsible calling of a teacher. Teaching is an *art* quite as difficult as any other, and for its practice certain qualifications are demanded which are far from being in the possession of all. The teacher must not only know *what* he has to communicate, but also *how* he ought to do it. Nor is this all; for if all instruction is in reality to be *education*, the teacher must rise from the instructor to the educator; he must not only understand how to impart knowledge and dexterity, but also how to impart both in such a manner that they make for the mental development of the pupil, especially with regard to moral training. But as we cannot give to others what we do not ourselves possess, it must necessarily follow that only he who is himself educated can have an educative influence over another. Therefore, exactly in proportion to the educative aim of the teacher does his *personality* enter as an important factor into the work of instruction. Now, since slöjd is to be regarded more as a means of education than a subject of instruction, in the common acceptation of the term, the first demand of all made upon the teacher who undertakes it must be that he should feel himself to be an educator, and strive without ceasing to improve himself as such. This, however, is not sufficient. To be a teacher of educational slöjd, it is necessary to be familiar with its aims, and with the means by which these are to be attained. *One* of these means is the possession of what is called *technical* dexterity, *i.e.*, dexterity in the right use of tools, and in the accurate production, by their means, of articles involving the exercises required by the particular kind of slöjd in question. The importance of this dexterity must neither be over-estimated nor undervalued. Unfortunately one or other of these errors is frequently committed. On the one hand it is maintained that if a person can only prove that he possesses technical dexterity in sufficient degree, *i.e.*, if he himself can *Technical dexterity.*

produce good work, he thereby fulfils one of the most important requirements of a good slöjd teacher. From this point of view the skilful artisan or "Slöjder" would be the best teacher of slöjd, because he can with justice be held to possess the best technical qualifications. Past experience, however, has shown that, as a rule, the skilful artisan or "slöjder" is not the best person to fill the responsible post of the slöjd teacher. This follows from the very nature of the case. The artisan has acquired his technical dexterity in a totally different way, and for a totally different purpose, from what is required in educational slöjd. Technical dexterity is the principal thing with him. It is before every other consideration a source of income. In educational slöjd, on the other hand, it is to be regarded only as one means among many whereby the teacher is able to bring an educative influence to bear on the pupils. The artisan who has great technical skill is too often tempted while teaching to use this skill in a way which may be for the advantage of the work with which the pupil is occupied, but is certainly not for the advantage of the pupil himself. His "instruction" consists not infrequently of work which he does for the pupil, with results which are excellent from the economical point of view, but which are very objectionable in their educational aspect. Partly for this reason and partly because the artisan often does not understand how to maintain really good discipline with children; and because, moreover, he is unacquainted with the general principles which apply to all instruction, it has been remarked, that where instruction in slöjd is concerned, even a very capable artisan often falls far behind the results attained by those who are in his opinion little more than bunglers, and who may be far inferior to him in technical dexterity. At the same time, it is by no means intended to convey the idea that the skilled artisan may not be a good teacher of slöjd—provided he understands the difference between slöjd and his trade, and is in possession of the other necessary qualifications—but it is maintained that in such a case it is less *because*

[T]he artisan [a]s a teacher [o]f slöjd.

he is an artisan than in spite of it, for the first condition is that he must renounce the traditions of his craft, and become penetrated by educational ideas.

But the truth here, as in so many other cases, lies between the two extremes. It is as hurtful to under-estimate technical skill as it is to over-estimate it. Therefore, let no teacher imagine that he can successfully undertake instruction in slöjd with slight and superficial knowledge on the purely technical side. It will soon and surely be made clear to him that this is not the case. If he has not himself the necessary technical dexterity for his purpose, it will be difficult, indeed almost impossible, for him to make clear to his pupils how they are to handle their tools and execute the work prescribed. Neither will he be able in an efficient way to supervise their work and criticise the quality of what they produce. The feeling of self-mastery which is so essential for the teacher when he stands face to face with his pupils, forsakes him, and the educative results which he intends to attain by means of slöjd are diminished in proportion. It is most important that this should be laid down once for all, because some teachers possibly imagine that the technical skill necessary for teaching may be obtained by attending one or two slöjd courses. This is by no means the case, and the organisers of such slöjd courses are the first to understand and to insist upon the fact, that they can only aim at laying a foundation on which students may afterwards build by means of independent work. Just as little as one can learn to play on any instrument by merely taking lessons for a given time from a music teacher, can skill in the management of tools be acquired and maintained without continuous and earnest practice. The teacher who feels real interest in slöjd must therefore, on his own account, endeavour to improve in respect of technical skill, and this will prove a two-fold gain, because the bodily exercise affords a healthy change from the mental work with which the time of the teacher is chiefly filled.

Under-estimation of the importance of technical dexterity.

To summarise what has been said in the foregoing: the teacher of educational slöjd must above all things have the habit of mind which is indispensable for the right performance of the teacher's work; his personality must be such as renders him fit to be a teacher; he must know the objects of educational slöjd and the means by which they are to be attained; and finally, he ought to have sufficient dexterity to handle the tools and to execute accurately the work which is incidental to the course of instruction. These are the demands made on him; may he strive to meet them.

The special kind of Slöjd recommended.

Various materials, *e.g.*, wax, clay, paper, pasteboard, wood, metal, &c., may be used in educational slöjd. Wood, however, is for several reasons the most suitable material; hence *wood-slöjd* has been the most popular of all, both in schools and for private instruction. As the name implies, wood-slöjd means "slöjding" in wood. This, again, includes several different kinds of work. Amongst these, however, it is the so-called *slöjd-carpentry* which best fulfils the conditions required when instruction in slöjd is given with educational ends in view. It is adapted to the mental and physical powers of children. By enabling them to make a number of generally useful articles, it awakens and sustains genuine interest. It encourages order and accuracy, and it is compatible with cleanliness and tidiness. Further, it cultivates the sense of form more completely than instruction in drawing does, and, like gymnastics and free play, it has a good influence upon the health of the body, and consequently upon that of the mind. Additional advantages are, that it is excellently adapted for methodical arrangement, comprising as it does a great number of exercises of varying degrees of difficulty, some of which are very easy; and that it gives a considerable degree of general dexterity by means of the many different tools and manual operations which it introduces.

löjd arpentry.

We must not confound slöjd-carpentry with the work done by the carpenter, properly so-called. This is the more necessary because great confusion of ideas prevails on the subject; not least, remarkably enough, amongst those who are interested in slöjd, or give instruction in it.

Difference between slöjd-carpentry and ordinary carpentry.

It must be borne in mind that although slöjd-carpentry and ordinary carpentry have something in common, inasmuch as the same raw material (wood) is employed, and to some extent the same or similar tools are used, yet they differ one from the other in several very important respects. For example, the articles made in slöjd-carpentry are in many cases quite different from those which fall within the province of the carpenter. The articles made in slöjd-carpentry are differentiated partly by their smaller size, for the articles made in workshops are generally much larger; partly by their form, for they are often bounded by variously curved outlines, whilst articles made by the carpenter are generally rectangular or cylindrical. This is especially shown in the case of the many different kinds of spoons, ladles, scoops, handles, &c., &c., which form such an important element in slöjd-carpentry.

Further, though many tools are common to both kinds of work, there are also considerable differences in this respect. Several tools which are seldom or never used in the carpenter's workshop, *e.g.*, the axe, the draw-knife, and the spoon-iron, occupy an important place in slöjd-carpentry.

The most characteristic tool in slöjd-carpentry is, however, the *knife*, and by the use of this, his chief instrument, the slöjder may always be distinguished from the carpenter, whose favourite tool is the chisel, and who, as seldom as possible, and never willingly, takes the knife in his hand. In carpentry, on the other hand, use is made of a number of tools more or less necessary, which are quite unknown to the slöjder, who works for the most part under more primitive conditions. Distinct differences can also be pointed out in the manner of executing the work (for while division of labour is practised in carpentry, it is not permitted in slöjd) and in the manner of using the tools. It will be seen from the foregoing that much may pass under the name of instruction in slöjd which, properly speaking, ought simply to be called instruction in carpentry. It is most important that this distinction should be maintained, because otherwise educational slöjd will by degrees be lost in instruction in carpentry as a trade.

Turning and wood-carving.

In some schools where slöjd is taught we find *turning* and *wood-carving* as well as slöjd-carpentry. This, however, is not so common now as it was a few years ago. People seem to be coming more and more to the conclusion that both occupations are more suitable for the home than for the school. Neither of them is to be commended from the hygienic point of view. As regards turning, the difficulty of procuring suitable turning-lathes presents in many schools a serious obstacle to its general use; whilst the necessity of performing preliminary exercises, apart from the actual objects made (a proceeding of very doubtful educational value) places turning quite in the shade as compared with slöjd-carpentry. Wood carving, on the other hand, does not involve that energetic bodily *labour* which is of such great importance in connection with educational slöjd. Again, wood-carving, classed as it is with the so-called "finer" kinds of manual work, has a tendency to intensify in the child that contempt for rough bodily labour which has already unfortunately done so much social harm. The danger of this is however greatest when the children are imprudently permitted to ornament objects which they have not made. When wood-carving is used, not as a separate kind of slöjd, but in order to complete slöjd-carpentry, and when ornamentation is only allowed *after* the children are able in a satisfactory way to execute the articles to be embellished by its means, the disadvantages are minimised.

Method.

The meaning of method.

Systematic action, directed towards an end, is termed method. Every form of human activity, in so far as it is concerned with the attainment of a definite preconceived end, must therefore be regulated according to method, and this universally applicable rule holds good in the case of that activity which is directed towards instruction and education. Hence great importance has always been attached to methods of instruction. In fact, in many cases too much attention has been paid to the study of special methods. Not that we agree with those who, by strange confusion of ideas, regard the rules of scientific method as opposed to practice, saying:—"We are practical people, and therefore we mean to teach in our own practical way, not to follow the theoretical methods of others." They thereby show that they do not understand how, in the very nature of things, there can be only one really practical mode of procedure, and that is the method which is in harmony with sound theory, and that any other way of going to work must be more or less unpractical. On the other hand, it cannot be denied that many teachers misunderstand the true significance of method to such a degree that it becomes the Alpha and Omega of the work. They forget that, strictly speaking, method is merely a *tool*—though a very necessary one—in the hand of the teacher; and that, just as little as a tool can execute a piece of work of its own accord, just so little can method ever be the chief factor in instruction. The teacher's power to apply method is the determining factor. A good method in the hands of a truly capable teacher will always give better results than a bad method. The best method is of comparatively little value if the teacher is inefficient.

It will now be clear that slöjd, whether regarded as a subject of school instruction in the usual sense, or as a purely disciplinary subject, must be treated according to rules of method. The ordinary rules of method can be applied to it; and chief amongst them those which are generally regarded as fundamental principles, namely, *that instruction shall proceed gradually from the more easy to the more difficult, from the simple to the complex, and from the known to the unknown*, it being always understood that the starting point is sufficiently easy, simple, and well-known.

The exercises. In drawing up a system of method in slöjd teaching it is difficult to find any fully logical principle of arrangement elsewhere than in the *exercises*. By exercises in this connection is to be understood that manipulation of the materials by means of one tool or more in a definite way, for a definite object. Now these exercises can be arranged in a series, in conformity with the rules given above. This could not be done so easily if the tools themselves constituted the principle of arrangement, because, *e.g.*, in the case of two tools, some exercises performed with the one may be easier, and some on the contrary may be more difficult, than the exercises which are performed with the other. It is obvious that the models cannot constitute the principle of arrangement, because they are merely the incidental expressions of the exercises. When, therefore, it is said that the models in a series are graded from the more easy to the more difficult, it is meant that the exercises occurring in these models proceed in this way. The exercises themselves are partly simple, partly complex: the latter consisting of two or more simple exercises in combination. The given number of exercises entering into the work of special kinds of slöjd depends more or less upon opinion, for it often happens that what is regarded as one exercise might be analysed into two or more, or might be considered as a part of a more complex exercise. Hence the eighty-eight exercises in slöjd-carpentry enumerated further on, might easily be increased or decreased in number, depending entirely upon how far it is considered advisable to carry this analysis or synthesis.

Method of teaching the exercises.

The exercises, their number, their names, and their order are not, however, the only factors which determine method in slöjd. The way in which they should be taught must be included. There are different modes of procedure. One of these is to teach the exercises one after the other, simply as isolated or "abstract" exercises, until they have all been performed. This may be justified from the point of view of method in general, but opinions may differ, not to put it too strongly, as to its educational soundness. Another mode of procedure is to apply each exercise, after it has been practised separately or in the abstract, in the construction of a given object or model. The exercises themselves are thus given as *preliminary practice.* This, though certainly a step in the right direction, does not fully satisfy the demands of educational method, which requires us to proceed from the concrete to the abstract, and not *vice versa;* and, moreover, such unnecessarily round-about methods cause the loss of valuable time which might be better employed. Method in slöjd only becomes educationally sound when the pupil, by constructing objects which can be used in everyday life, acquires dexterity in performing the exercises as they occur. To take an illustration from language teaching, the first mode of procedure corresponds to the learning of abstractions in the form of grammatical rules; the second corresponds to the application of these rules in sentences after they have been learned; the third corresponds to the method by which the pupil is led up, through sentences or combinations of sentences, to the laws of language which in them find expression.

Arrangement of a series of models.

There are, however, other fundamental principles which must be adhered to in arranging a series of models in such a way that the exercises involved shall follow each other in methodical order. The general nature of the models and the manner in which the exercises ought to be introduced in them must be considered. In choosing a series of models the best plan is undoubtedly to consider local conditions, and endeavour to make it exactly representative of articles which can be used in the homes of the pupils. By this means interest in the instruction given is better aroused and maintained, not only in the pupils, but—and this is quite as important—in the parents, and thus the bond between the school and the home is strengthened.* Opinion is now probably almost unanimous that all articles of luxury should be excluded. (Such articles, however, are by no means synonymous with articles intrinsically beautiful.) The interest of the pupils is also heightened if the first articles presented to them are no larger or more difficult than can be executed satisfactorily in a comparatively short time. The first models ought, on this account, to include few exercises; and it may be laid down as a general rule that, as far as possible, each successive model should include only one new exercise, or two at the most. In the arrangement of the series, attention must also be paid to alternation in the form of the models. The articles which are included in slöjd-carpentry consist partly of "*modelled*" *articles* bounded by curved surfaces, and partly of *rectangular articles* bounded principally by plane surfaces. It is very important that any arrangement of models in a series should present good alternation between these two kinds, and, generally speaking, a modelled object should follow a rectangular object, and *vice versa*. As a result, each model acquires to some extent the

* As some confusion of ideas appears to prevail in England between the importance of the *educational principles* on which slöjd is based, and the *models* in which these principles are exemplified, it seems desirable to draw the attention of readers to this passage. It indicates sufficiently clearly that, in whatever country Swedish slöjd may be adopted, the more familiar and the more serviceable the articles made are to the inhabitants of that country, the more nearly will the method of teaching conform to one of the great principles of educational slöjd, viz.: that the pupil's interest shall be excited and sustained by the making of articles which he himself or the other members of his family can use. Many of the models at Nääs have, within the last year or two, been either modified or changed entirely in order to render them suitable for English students, and it is incumbent upon every slöjd teacher to make his own series of models conform to the ideas and requirements of the people among whom he teaches, keeping in view the general principles of method which would apply to any series.—TRS.

charm of novelty, and this still further increases in the pupils that interest for their work which is of the very greatest importance as regards the educational benefits to be derived from slöjd.

Intuitional nature of the instruction.

The manner in which the details and finished appearance of the objects he is to execute are made clear to the pupil, must be included within the province of method. It is assumed that in this, as in all other instruction, it is of the highest importance that the teacher strives to make his teaching as *intuitional* as possible. To this end, in the elementary stages, the models should always be executed after drawings and models, and in the first instance invariably after *models* which are placed before the pupils for accurate imitation.

As, however, it has been proved to be difficult, in many cases indeed almost impossible, to preserve even a well-executed wooden model in its original shape and size, and as, for other reasons, it is highly advantageous to connect instruction in slöjd with instruction in drawing, the model should be copied to as great an extent as possible by the aid of geometrical constructions, sufficiently simple to require in the pupil only a slight acquaintance with geometrical drawing. In addition to this the most important measurements of the model's dimensions should be given, in order that the pupil may make use of his rule or metre-measure.* By degrees drawings in perspective and projections may be introduced as patterns *together with* the model; and finally, when the pupil has reached the highest stage, and has attained sufficient dexterity in slöjd and in the interpretation of a drawing, the model may even be taken away, and the work executed

* As the metrical system of measurement admits of greater exactness than our English system, and as it seems desirable to accustom English children to its use, teachers of slöjd are strongly advised to adopt it in connection with the dimensions of the models. No difficulty need be anticipated. It has been found that, in cases where children were permitted to use either their English foot-rule or the metre-measure, they invariably preferred the latter.—Trs.

after a drawing only. This may be regarded as the *final aim* in elementary instruction in slöjd.

Self-reliance. It is an essential condition of any method of instruction in *educational* slöjd, that the work of the pupils shall be *independently* and *accurately* executed, for only thus can habits of self-reliance, order, and accuracy, so important in the formation of character, be developed. In order that self-reliance may be developed, the teacher must guard himself against giving more help than is absolutely necessary, whether this help consists in explaining the best way of doing the work, or in doing the work instead of the pupil. As regards the latter, the teacher will do well to lay down, as a general rule, that he *never* should touch the pupil's work, for only by this means can he avoid the temptation, to which unfortunately many teachers have succumbed, to execute the most important parts of the work instead of the pupil. At the same time he must remember that it is also hurtful to the pupil, and that it deprives his instruction of considerable educational value, if by unnecessary explanations he hinders the pupil from using his own judgment to discover the right way. The teacher's art in educational slöjd consists essentially in being as passive and unobtrusive as possible, while the pupil is actively exercising both head and hand. Only in this way can the feeling of self-reliance arise and gain strength. Let the teacher content himself with pointing out the way, and watching that the pupil walks in it. Let him as much as possible refrain from leading where this is unnecessary and, it may be, hurtful.

Accuracy. In order to develop the habit of accuracy in the pupil by means of slöjd, it is essential that he should make his model as nearly as possible an exact likeness of his pattern, or—when the model has changed in shape and size—an exact copy of what it ought to be, as indicated by the geometrical construction, or complete drawing and given measurements. We very often hear people say that it is quite unnecessary to be so particular with the work, since *e.g.*, a flower-stick can be quite as serviceable whether it is a little shorter or a little longer. This is perfectly true on the assumption that the making of a serviceable flower-stick is our chief end in making it. In educational slöjd, however, the principal object is not the article made, but the mental and physical benefits which accrue to the pupil by means of the work. In this case it cannot be unimportant that he should be exercised in the endeavour to execute something as well and as accurately as he is able to do it. For in this way his natural disposition to work carelessly is checked, while at the same time the degree of accuracy to which he is gradually accustomed will be of great advantage farther on in the series of models, when he has to perform such operations as mortising, grooving, dovetailing, &c., which call for no inconsiderable degree of accuracy in their performance. Though a pen-holder need not be of any exact size, this is by no means the case with the joints in dovetailing; and in making the former exact, the latter operation is rendered possible, or at all events easier. At the same time, we must not demand of the pupil work which is absolutely correct in all its details, for this clearly lies beyond his powers. The teacher must exercise his "tact" as an educator in determining the degree of accuracy which is to be demanded of every separate pupil in every separate model, and this being done, the teacher must unhesitatingly reject the articles which fail to come up to the required standard. But in order that the pupil may not be disheartened by repeated rejections, it is advisable not to insist on the repetition of the same model more than, at the outside, *three consecutive times*. If the pupil fails to succeed the third time, he should be allowed to pass on to the next model, and not required to return to the one he failed to make, until he has succeeded in making the other; this he usually does easily enough, owing to the increased facility he has gained by practice. If the pupil is permitted to pass over a model altogether without bringing it up to the required standard, it may encourage him in caprice, and counteract the development of habits of perseverance, the acquisition of which is of such great importance in life. Further, the general rule should be strictly observed that every article is to be executed as well and as beautifully as possible. In educational slöjd it is much more important that what is made should be the product of good and conscientious labour, than that much should be produced. Therefore, whatever bears the impress of carelessness and haste must be rejected without mercy, lest the pupil fall into bad habits, and the educative influences of slöjd be weakened.

Individual instruction, versus class-teaching.

The question whether individual instruction or class-teaching should be adopted, comes also under the head of method in slöjd-teaching. As the aim in educational slöjd is totally different from mere mechanical instruction in the art of using tools and making articles, it may be laid down as a principle, that only in the degree in which the personal influence of the teacher reaches each individual pupil, can his influence be truly educative. And as human beings differ greatly from one another in natural disposition and other respects, instruction, in order to reach the highest degree of educative value, must be specially adapted to each individual. It is as easy to explain, point out, lead, and help *too much* as *too little*, and thus to check that mental development which can only be secured by systematic well-balanced effort. This is, and this will continue to be, the disadvantage of class-teaching:—this term being assumed to mean, instruction during which all the pupils taking part in the lesson have their attention directed at the same time to the same part of the subject. This disadvantage can never be lost sight of, but in the case of several subjects of instruction, especially the purely intellectual subjects, it is counterbalanced to some degree, because, by means of class-teaching, the practical benefit is gained that a teacher can teach a larger number of pupils than he could teach individually. Slöjd, however, does not belong to these subjects, because in it the teacher's powers are limited, to start with.

by the number of pupils he can efficiently *supervise* at work; and it can speedily be demonstrated that he cannot, in class teaching, supervise more than by individual instruction, *provided that in each case equally good results are aimed at.* On the contrary, he may find that he cannot supervise so many. Another practical objection to class-teaching in cases where slöjd is applied to *educational* purposes, is the impossibility of keeping the class together in the execution of their work. It follows either that the more backward pupils scamp their work or are allowed to pass over some of the models in the series, or else that the superior pupils are checked in their progress, and thereby prevented from doing as many exercises as they otherwise could have accomplished.*

The leading question of method in educational slöjd teaching ought to be less *how much*, or *how many*, as *how well.*

The Pupils.

The age during which instruction can be received with advantage in any subject whatever is limited downwards as well as upwards by the work it involves. As regards slöjd-carpentry, children ought to have attained the degree of development which corresponds roughly to 10 or 11 years. Otherwise they cannot be expected to meet the demands made on the spirit of self-reliance during work. At the same time, as children of the same age differ greatly in point of development our guiding principle should not be the date of birth, but the mental and physical powers which the child has at command. What one child of nine years can accomplish with ease may be beyond the powers of another child of twelve. As regards the *upward* limit of age, it lies considerably beyond school years. *The age of the pupils.*

* On certain occasions it is advantageous to demand the attention of all the pupils at one time, *e.g.*, when the teacher wishes to explain the properties of a tool and the method of using it, or wishes to examine all the pupils together. These, however, are special cases, which ought to be quite independent of the slöjd-work itself.

The number of the pupils. The number of pupils who can be managed individually by one teacher at the same time varies considerably, and is influenced partly by the teacher's general efficiency, partly by his special efficiency, and partly by the stage at which the pupils are. The teacher who is unaccustomed to teach slöjd will probably be unable at first to manage with ease more than from 6 to 8 pupils, especially if they are beginners; later on the number may be increased to 12, and by degrees, under favourable conditions, to 15, 18, or at most 20.* No teacher, however, ought to let his desire to increase the number of his pupils induce him to take more at one time than he can manage in a thoroughly satisfactory way.

The Time given to Instruction.

The length and the distribution of the lessons. A slöjd lesson ought not to last less than an hour and a half, or more than two hours and a half. It ought, if possible, to intervene between hours devoted to intellectual instruction, because it offers a wholesome variety for mind and body. Slöjd by artificial light should be avoided as much as possible. It is desirable that every pupil should receive three lessons a week. They should be given every other day, and if the pupils have gymnastics on the intervening days, it will secure, to some extent at least, the necessary physical exercise on a rational basis.

The Slöjd-room.

The use of the school-room for slöjd. The days are past, in Sweden at least, when it was regarded as a degradation of the rooms devoted to intellectual instruction to use them for slöjd teaching. Since "practical" slöjd has been forced to make way for educational slöjd, and the importance of the latter has been more and more recognised, no thoughtful teacher can think that his class-room loses dignity because manual labour is carried on in it. In many schools where space is limited, slöjd must, at first at least, be

* This has been proved by observations made in the elementary schools in Stockholm.

given a share of the school-room. Where this room is large enough, and where the slöjd-teacher's spirit of order is sufficiently strong to make him keep his department always tidy, this combination may be made without special inconvenience. It is advisable to place the slöjd-benches and the tool-cupboard at one end of the room. The removal of desks to make temporary room for benches should only be permitted when such an arrangement is unavoidable.

The use of the school-room for the double purpose of intellectual work and slöjd is not, however, to be recommended when circumstances permit of separate rooms being fitted up. Different arrangements are required for the two branches of instruction. A description of the general arrangement required in a room devoted to the purposes of educational slöjd carpentry follows. This description is based on experience gained in the teaching of slöjd up to the present time. It must, however, be borne in mind in this connection, that the conditions in an elementary school in the country and in a school in a large town or in closely-populated manufacturing districts, vary according to circumstances.

The slöjd-room in ordinary country schools.

As regards the former, we must, as a general rule, be less exacting in our demands; in the latter, on the contrary, arrangements may be made which shall meet fully the educational requirements of a good slöjd-room. In an ordinary elementary school in the country, where there may not be more than from eight to twelve pupils requiring instruction at the same time, a slöjd-room measuring 16 ft. in length, 13 ft. in breadth, and 10 ft. in height, will be large enough. It should be situated on the ground-floor. The walls should be wainscotted, and the room should contain three or four double, or six or eight single, benches; cupboards for tools, models, and finished articles; a grindstone, a chopping-block, and, if turnery is included in the course, also a turning-lathe. If the room is kept locked between lessons, the tools may be disposed round the walls instead of being kept in a cupboard. The wood required should be stored in some place adjoining.

'he slöjd-
oom in
arge
chools.

In a large school, where opportunity is given for making the arrangements for slöjd teaching as complete as possible, the following directions may be found useful:—

Situation.—The slöjd-room (not "work-shop") should open from a lobby either on the ground-floor or on the top storey. It should never be situated in the basement. If it is on the ground floor, care should be taken that it is as far as possible from the other school-rooms, that the noise may not disturb the pupils in the latter. If a slöjd-room is situated *above* a school-room, it should be furnished with a double floor, with an intervening layer of sawdust to deaden the noise.

Area.—To accommodate 20 pupils at one time, the room should be about 50 ft. long and 23 ft. broad. This will give adequate space for 20 separate benches (placed in 5 rows), a turning lathe, a saw-bench, grindstones, chopping-blocks, cupboards for tools, models, and finished articles, and a rack for wood. Wood ought not to be suspended from the roof if this can be avoided, partly because it is unsightly, and partly because it gives unnecessary trouble. The space between the benches ought to be about 2½ ft.

Height.—The slöjd-room ought to be from 12 ft. to 15 ft. high.

Windows.—The slöjd-room should be well-lighted by large and properly placed windows. The area of window surface is generally reckoned as 25 to 30 per cent. of the floor area. If the slöjd-room is on the ground floor, windows should, if possible, be placed in three of its walls. If it is on the top storey it is better to let the light enter by sky-lights than dormer windows.

The Walls.—To prevent injury to the walls they should be lined with wood, or at all events with a tolerably high wainscotting. The doors and the window-frames should be painted, but the walls need only be varnished.

Warming.—Where there is no central system of heating,

the best way to heat the slöjd-room in winter is by two large stoves. The temperature ought not to be higher than from 54° to 57° F. The glue should be melted on a small stove heated by gas or oil.

Artificial light.—As work is done in the slöjd room on winter afternoons, arrangements must be made for artificial light. This light, whether furnished by gas or by electricity, must always come from above, in order that no shadows may be cast on the work.

The Position of the Body during Work.

If slöjd is to contribute towards physical development—a point on which most people are now tolerably unanimous—methodical and effectual arrangements must be made to this end.

Slöjd and gymnastics should go hand in hand.

It can easily be demonstrated that of all kinds of slöjd, slöjd-carpentry in conjunction with gymnastics is the best adapted for physical training, but it is equally clear that this can only be the case, provided that the positions assumed and the motions prescribed are well-selected. As regards this we have to find the happy medium. On the one hand it cannot be denied that many of the so-called "instinctive" positions assumed by artisans and "slöjders" have reference more to what is advantageous for the work than for the worker. On the other hand it must be granted that though slöjd ought to be considered as "applied gymnastics," this principle should not be carried out so pedantically that the idea of work is lost sight of. Slöjd is essentially *work,* and not merely gymnastic exercises with tools as apparatus; and all that we are justified in aiming at is, that when we have a choice between positions and movements favourable to physical development and those which are unfavourable, we must adopt the former. We may rest assured that, in the long run, not only the worker but the work will gain thereby.

Slöjd is work, not merely gymnastic exercises.

Harmonious or all-round physical development is mate-

Uniform exercise of the muscles of both sides.

rially advanced when the muscles of both sides are equally exercised during work. This is a fundamental rule in gymnastics. It is equally binding in slöjd whenever it is capable of application. The objection we sometimes hear that the left hand has not the same strength and steadiness as the right, depends on a confusion between cause and effect, because this inferiority in most cases is caused by the fact that at an early age the left hand in the matter of exercise is neglected for the right. It is, moreover, easy to enumerate a great number of operations in which both hands execute almost the same work. As examples may be given: sowing seed, kneading dough, weaving, hewing wood, driving, rowing, playing on the piano, &c. In slöjd-carpentry the *saw*, the *plane*, the *centre-bit*, and the *file* may, in particular, be directed alternately by the right and by the left hand, and the change should be made by all the pupils together, at the command of the teacher, about every half hour. On the other hand, the use of the axe or the knife by the left hand is not to be recommended until great experience in the use of the left has been gained, on account of the greater danger of injury should the tool accidentally slip aside.

Positions and movements during work.

The following general rules may be given for the positions and movements in educational slöjd-carpentry.

Position of the chest.—The chest encloses the important vital organs, the heart and the lungs, the former of which regulates the circulation of the blood, and the latter the process of respiration. That these may freely and without hindrance perform their functions, the space in which they move must not be diminished. It must rather be enlarged. We must therefore endeavour to prevent any narrowing of the chest, and attention should always be directed to keeping the shoulders well back during work, in order that the chest may be expanded. Inspiration and expiration should take place quietly, without any effort whatever.

The head should be held as erect as possible, to avoid unnecessary loss of muscular power, to permit greater freedom of circulation, and to preserve the eyesight from injury during work. When the head is bent forwards the veins in some situations are compressed, in others extended; in both cases their calibre is diminished. In connection with the effect the position of the head may have upon the circulation, the importance of loose clothing should be noticed. Tightly fitting collars and neckties should be above all avoided. To preserve the sight, work should not be held nearer the eye than about 12 in.: for this reason it is very advantageous in educational slöjd to use exclusively benches whose construction permits of their being raised to different heights. Thus the work may always be held at the proper distance from the eye, while the position of the head is, from the hygienic point of view, most advantageous.

The feet should be so placed as to afford the best and firmest support during work. In the execution of every exercise a certain mechanical resistance has to be overcome. For this purpose muscular strength, and in certain circumstances the weight of the body, must be called into play. This resistance must be regarded as force opposing the worker in a certain direction, and he must allow his body to assume the state of equilibrium most favourable in relation to the direction of the force. This is done as regards the feet, when the line of most resistance is in front of the worker, by placing the one foot in front of the other in such a position that a line drawn from the foremost foot in the direction of its length, would meet the heel of the other at right angles; and when the resistance is from the side, by placing the feet apart sideways. A bad habit of frequent occurrence, especially in planing, is to turn the toes in. This ought to be avoided as much as possible, because it interferes with the natural action of the knee joint.

The position and movements of the body.—The worker should assume a position, in relation to his work, which enables the muscles of his arms to have free play in the most favourable direction for its execution, *i.e.*, in a direction opposed to the line of resistance, or friction between the tool and the piece of wood. In certain exercises, such as planing and boring, this friction is, to some extent, increased by the necessary bending of the body over the tool, whereby the weight of the body helps to press it against the wood. In using some tools, *e.g.*, the saw, this weight may also act as a kind of regulator, by gently setting the body in motion backwards and forwards. The reader is referred to *Plates I.-VIII.* for illustrations of some of the most important positions.*

Some Rules for the Slöjd Teacher.

Order an indispensable condition.

In all teaching, and not least in slöjd teaching, the maintenance of order must be laid down as an indispensable condition. The following simple directions may serve for guidance to the teacher.

The pupils' places for work.

Every pupil should have a fixed place at a bench. When circumstances permit, it is advisable to have at disposal as many benches (or when benches intended for two are used, half as many benches) as there are pupils taking part simultaneously in a lesson.

Numbers on the benches and tools.

The benches and tools should be furnished with numbers, so that they can easily be distinguished from one another. The following tools should, if possible, belong to each bench, and be marked with its number: knife, trying-plane, smoothing-plane, jack-plane, square, marking-gauge, compasses, rule or metre measure, and scraper.† Other tools may serve the whole class in common.

Fixed places for tools.

All tools should have fixed places. Those belonging to the bench may be allowed to lie upon it until the close of the lesson, but all tools in common use should be laid by or hung up immediately after use, in order that they may be easily found.

The teacher must take care that all the edge tools in use are

---

* These plates are specially intended to illustrate the position of body which the worker should assume when beginning the particular exercise indicated.

† These constitute the bench-set.

well sharpened, and that any tool which gets out of order, or is broken, is repaired as soon as possible. If practicable, the pupils should do their own repairs. *The sharpening and repairing of tools.*

At the beginning of the lesson the pupils should, in an orderly way, get out their tools and work. The latter, if begun in a previous lesson, should be kept in boxes specially provided for the purpose, and should be marked with the pupils' names. *The work.*

In order to teach and superintend in the full meaning of these terms, the teacher must not stand still in one place. He must go from one pupil to another with advice and criticism. The pupils, on the contrary, must, as far as possible, remain at their benches. If they desire any advice from the teacher, they must not attract his attention by calling out, but by some signal, *e.g.*, holding up one hand, standing in front of the bench and looking towards him, etc. All unnecessary talking must be carefully avoided. *Teacher and pupils during work.*

The pupil himself, guided by the teacher, must select suitable wood. Waste must be avoided as far as possible. *Selection of wood.*

The pupil must not be allowed to polish with sand-paper until the teacher has examined the work and found that sufficient use has been made of cutting tools. The sand-paper is to be kept by the teacher and given out by him as required. About 6 sq. in. is calculated for each model. The calculation is founded on the supposition that though the models become larger as the course proceeds, the greater facility of the pupil diminishes in about the same degree his need of sand-paper. *Sand-paper.*

At the end of the lesson all the tools should be put back in their places, care being taken that all the saws are loosened. The tools should be counted by the "captain," or monitor, appointed for the class, after which the teacher sees that everything is in its right place. The wood and the pieces of work are put away tidily. The benches are brushed and made clean with a brush which should hang by the side of each bench, and the floor is swept. The shavings, however, need not be carried away oftener than once or twice a week. *Putting the slöjd room in order.*

Finished work.

When the finished pieces of work have been "passed" by the teacher, a label should be stuck on, and on this label should be stated the number of the model and its name, the name and age of the pupil, and the number of hours spent in making it. If it is considered desirable to give every piece of work a value, this also may be mentioned on the label.

Taking the work home.

Although from the educational point of view it is advisable that the pupils should at once take home their work, it is generally for other reasons more expedient that it should remain in the school in the care of the teacher until it can be exhibited publicly at an examination or terminal breaking-up. After this has taken place, the articles are to be regarded as the property of the makers. The sale of work for the benefit of the school should never be thought of.

A very good plan is to allow the pupils to take home their work as soon as it is finished, in order to show it to their parents, on the understanding that, after they have seen it, it is brought back to the school, to be kept there as long as necessary.

Daybook.

The teacher should enter in a day-book, arranged for the purpose, careful notes regarding the pupils taking part in the slöjd lessons, their presence and absence from lessons, the articles they make, etc., etc.

CHAPTER II.

Wood, or Timber.

The material generally used in slöjd-teaching, and most suitable for the purpose, is *wood* or *timber*.

Intelligent knowledge of the material used is as essential to the teacher as acquaintance with the tools required. He ought, *e.g.*, to be familiar with the qualities which render different kinds of wood more or less appropriate for different purposes. Accordingly the description of the tools given in Chapter III. is here preceded by a brief account of the growth of trees; of the most important properties of wood, and the principal changes which it undergoes; and by a comparison of the technical qualities of the various kinds of wood in common use.

A. The Structure and Composition of Wood.

Wood or timber forms the greater part of the stems and branches of trees and shrubs.

To examine the inner structure of a tree-stem, a section may be made at right angles to the direction of its length, *i.e.*, a transverse or *cross* section; or from the pith to the bark in the line of one of the radii and parallel to the direction of the length, *i.e.*, a radial section; or a third section may be made at right angles to both the preceding as a tangent to the circumference, *i.e.*, a tangential section. *A tree-stem in section.*

On examining the cross-section of a stem we find an outer ring, the *bark*, consisting of a *corky layer*, the outer bark, and the inner bark or *bast*; next comes the *wood*, constituting the chief portion of the stem, and in the central part is a canal, filled with soft cellular tissue called the *pith* or *medulla*. Between the wood and the bast lies a narrow, light-coloured ring, the *Cambium*. This consists of a layer of embryonic cells, from which are developed on the one side wood, and on the other bast, and it is *here* that the growth of the tree takes place.

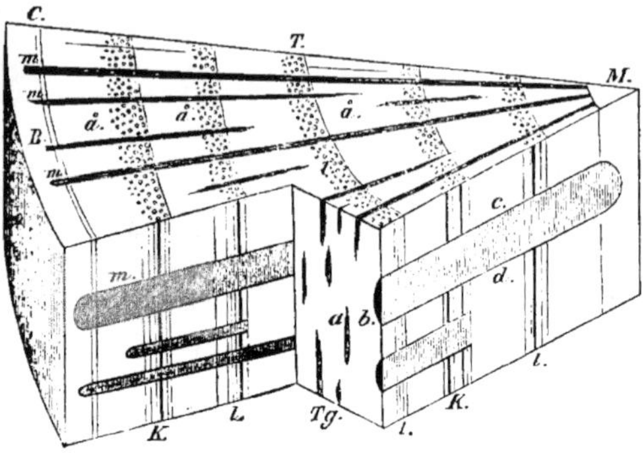

Fig. 1. Three sections of a tree-stem, at right angles to one another.
T. cross section, *K.* radial section, *Tg.* tangential section. *M.* medulla or pith, *B.* Bark, *C.* Cambium, *aa.* Concentric annual layers, *mm.* Medullary rays, *ab.* thickness of medullary rays, *cd.* height of medullary rays, *ll.* vessels.

The Cambium forms the soft, moist, spongy mass which may be seen under the bark in spring when the sap begins to rise. It consists of microscopic cells, some of which are long, prismatic, and pointed at the ends, while others are shorter and have ends which terminate abruptly. The inner bark and wood are developed chiefly from the long cells, the medullary rays from the short ones.

Wood Cells.

The formation of wood. The young cells from which wood is developed have at first very thin walls. They are filled with *sap*, the fluid which nourishes the growing tree, and which circulates with ease from one thin walled cell to another, and thus permeates the whole of the tissue. Gradually the walls of the cells become thicker; the cell contents solidify; the sap flows less and less freely; the whole tissue assumes the characteristics of wood, and ceases to take part in the circulation and assimilation of the sap.

The cellular tissue consists chiefly of *cellulose,* the chemical constituents of which are carbon, hydrogen, and oxygen.

Wood Fibres.

The cells from which wood is developed are principally the long-pointed cells. They lie close together and overlap one another at the ends, thus forming minute *tubes* or *fibres.* The zone of wood in any stem consists of these fibres massed together, and extending in the direction of the length of the stem. The connection between separate fibres is often very slight, as is shown by the ease with which they may be separated.

In trees of regular growth the fibres are straight and parallel. Wood of this kind is called "straight fibred." It is easily split. This is not the case with wood in which the fibres are crooked, or twisted about one another, as in gnarled or mis-shapen trees. The fibres in the root, the lower part of the stem, knotty branches and rough excrescences are always crooked, and sometimes they are twisted and involved in the most remarkable way. This gives rise to the peculiar speckled and veined appearance which is so highly prized in some kinds of wood.

The bast also consists of fibres, but they are longer and usually tougher than wood-fibres.

Concentric Annual Layers.

A new layer of bast and a new layer of wood are formed annually. This new formation goes on rapidly in spring and early summer, when vital activity in the tree is at its height. The cells are then large, and the wood formed from them, *i.e., spring wood,* is soft and loose in texture and light in colour. After the tree has budded the formation of wood goes on for a time, but less actively. The cells diminish in size and in diameter, and are more closely packed together. The wood formed at this period—*autumn wood*—is generally darker in colour and closer in texture than spring wood. There are fewer vessels (see p. 31) in autumn wood; in spring wood, on the contrary, they are numerous and quite visible as pores.

Distinctness of the concentric annual layers.

In consequence of the characteristics of autumn wood, the boundary line between two periods of vegetation is clearly defined, and it is easy to distinguish the *concentric annual layers* which mark each yearly increase in growth.

These layers are most sharply defined in needle-leaved trees and in some broad-leaved trees, *e.g.*, the oak, the ash, and the elm.* They are less conspicuous in the birch, the aspen, the alder, etc., and in some cases it is even difficult to distinguish them at all. As a new layer of wood is formed every year, the age of a tree may be determined by the number of layers.

Determination of the age of a tree.

In the tropics, where vegetation goes on during almost the entire year without any well-marked period of rest, the concentric annual layers disappear entirely.

The breadth of the concentric layers varies in different trees. In some cases they are more than 1 inch broad, in others scarcely $\frac{1}{60}$ inch. Their breadth may vary even in the same stem, depending on the more or less favourable weather of successive seasons. The layers on the side exposed to the south are often broader than those on the north. In old needle-leaved trees we usually find very narrow layers nearest the pith; beyond these the layers widen for the greater portion of the stem, and then contract once more until the outermost ones are often so narrow that they can with difficulty be distinguished by the naked eye. See Fig. 2.

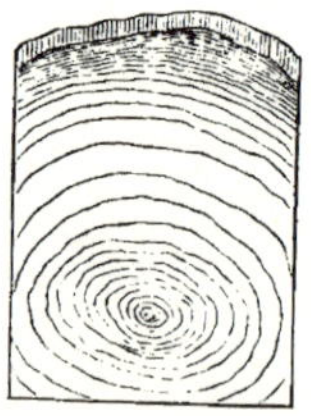

Fig. 2. — Showing manner of growth in needle-leaved trees.

* The terms *needle-leaved trees* and *broad-leaved trees* used throughout this book may be taken as practically synonymous with Conifers and Dicotyledonous trees.—Trs.

Narrow annual layers betoken good wood in needle-leaved trees; but the opposite holds good in the case of broad-leaved trees with large pores, *e.g.*, the oak, the ash, and the elm. Here broad annual layers are characteristic of a good quality of wood, because the pores which render the wood open in the grain occur chiefly in that portion of the layer which is formed in early spring, and are less numerous in the closer tissue of the autumn wood. See Fig. 3.

Close and loose timber.

Fig. 3.

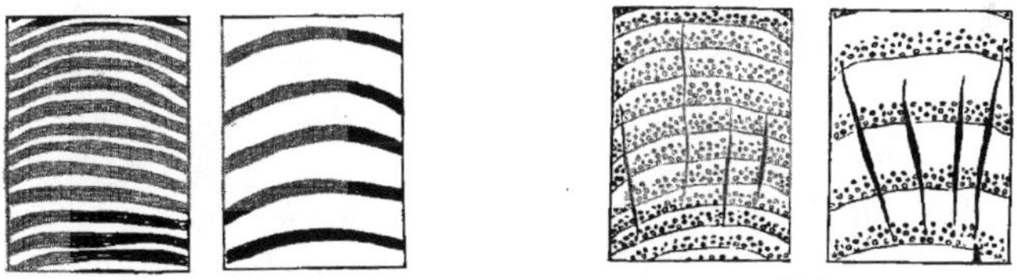

Fir.

Narrow layers, hard resinous timber. Broad layers, loose fibred timber.

Oak.

Narrow layers, loose fibred porous timber. Broad layers hard timber.

Vessels or Air-tubes.

When a cross-section of a stem is carefully examined a number of minute holes or *pores* are seen. These are the mouths of vessels or air-tubes, which penetrate the whole substance of the wood, parallel with the fibres. Their function is to enable the air to circulate in the stem, and they are found even in wood of the closest grain, rendering it porous. Vessels are most numerous in the wood formed early in spring, and very few are found in autumn wood, a circumstance which helps to make the annual layers more distinct. According to the size of these vessels wood is said to be *fine* or *coarse-grained*.

The porousness of wood.

Each kind of tree has something peculiar to itself in the manner of distribution, the number, and the size of its vessels. They are most marked in the oak, the ash, and the elm, giving to the wood of these trees, when seen in vertical section, its striped or streaked appearance. In a number of trees on the other hand, *e.g.*, the birch, the vessels are hardly visible, and they are distributed pretty equally over the concentric annual layers, making it difficult to distinguish consecutive layers.

esin.

Needle-leaved trees have no air vessels, but have channels filled with resin, *i.e.*, *resin-canals*. These occur chiefly in the autumn wood, to which they give a darker colour.

Heart-wood and Sap-wood.

In many kinds of trees, when the stem is sawn across, a considerable difference may be observed between the appearance of the inner and older, and the outer and younger concentric annual layers. The inner layers are usually firmer and closer in texture and darker in colour than the outer, which are less compact, lighter in colour, and full of sap.

'he Heart-wood, the aluable art of the 'em.

The firmer, darker wood is called *heart-wood* or *duramen;* the looser, lighter wood, *sap-wood* or *alburnum*. As a rule the latter forms a comparatively narrow ring round the former, which constitutes the greater portion of the stem, and which, when sound, is the valuable portion on account of its firmer texture and greater durability.

The proportion which the heart-wood bears to the sap-wood varies in different kinds of trees. For example, in the case of broad-leaved trees, the proportion is largest in the oak, the ash, and the elm; least in the birch, the maple, the alder, the hornbeam, etc. In needle-leaved trees, it is greatest in the larch and the fir; least in the pine. The resin in these trees is found chiefly in the heart-wood. It greatly increases its closeness and durability, and darkens its colour.

The most striking example of the *difference* in appearance between heart-wood and sap-wood is presented by ebony, in which the former is black and the latter white.

The Pith and the Medullary Rays.

The *pith* forms a column in the central part of the stem, and the *medullary rays* radiate from the pith towards the bark.

The pith is looser in texture, and is composed of shorter cells than the wood. The shape and size of the column vary considerably in different trees. In some, *e.g.*, the yew, it is very thin; in others, *e.g.*, the elder, it occupies a considerable space.

The *medullary rays* or "*transverse septa*" are composed of flat cellular tissue, which forms thin vertical plates radiating towards the bark. During the first year of the growth of the tree, these rays originate in the pith, divide the patches of wood and bast, and reach as far as the bark. In subsequent years they are formed in connection with the new wood, not with the pith, and they extend into the bark. The medullary rays are the medium by which the pith and the wood are brought into communication with the bark. They also divide the wood into wedge-shaped bundles. They are seldom so straight and regularly disposed as is represented in the diagram (Fig. 1), but are generally more or less curved, and they often branch out obliquely. They vary considerably both in number and appearance in different trees, and thus, like the vessels, they serve as a guide to the recognition of different kinds of wood. For example, *oak* is easily known by the smoothness and glossiness of its broad medullary rays when these are seen in radial section. This gives to oak timber the beautiful figured appearance called "silver grain." The beech has also long, broad medullary rays. The maple is distinguished by the fineness and number of its medullary rays.

Different kinds of wood known by the character of the medullary rays.

In the greater number of loose-fibred, broad-leaved trees, the rays are very narrow, and scarcely distinguishable by the naked eye. This is also the case with needle-leaved trees, the rays of which are extremely numerous.

The cleavage of wood. The medullary rays affect to a considerable extent the ease or difficulty with which wood may be split. As a general rule, timber is easily split if it has broad rays like the oak and the beech, or if the rays, though numerous, are straight and narrow like those of the fir and the pine. Other circumstances, however, may determine the greater or less resistance which any given timber presents to cleavage.

The Sap.

Next to the wood the sap is the most important element in timber. Its chief constituent is water, which holds in solution various organic and inorganic substances, but its composition undergoes changes in the course of circulation through the different parts of the tree.

The sap materials are absorbed by the roots, and as *crude*, or *ascending sap*, are carried by the still active cells of the sap-wood to the leaves. Here, through the influence of light and air, the crude sap is changed and made fit for the nourishment and growth of the tree, and is called *elaborated sap*. From the leaves it descends in the bast tubes to the cambium, where the new wood and bast are formed.

The organic constituents of the sap. Amongst the organic substances which the sap holds in solution may be named, starch, sugar, colouring matter, tannic acid, and albuminoids. The latter render it very liable to fermentation, and when this takes place the wood decays. This is the reason why timber, felled when the sap is circulating, and allowed to lie unbarked, readily becomes "sour." It also explains why sap-wood decays more quickly than heart-wood.

When wood is burnt the *inorganic* constituents remain in the ashes.

Sap also contains substances which are not required for the growth of the tree, but which occupy space and channels in the wood. Amongst these substances are the volatile oils, which are found chiefly in needle-leaved trees, and of which *turpentine* is the most important. The *resin* or *gum* found in needle-leaved trees is also formed from these oils. *Tannic acid* is found in a great many trees, especially in the bark. It is known by its acrid taste, and it abounds chiefly in the oak, the fir, and the alder. When fresh timber in which there is a great deal of tannic acid is split or sawn, the acid makes the polished edge of the tool become blue-black in colour. *Turpentine, resin, gum, and tannic acid.*

The destructive effect of the albuminoids of the sap is counteracted by the turpentine, resin, and tannic acid.

Water Capacity.

The sap, as stated above, consists chiefly of water; and, as it circulates in the sap-wood, it follows that the latter contains more water than the heart-wood, and more in spring than in the height of summer. As a general rule the water contained in unseasoned wood is about 40 to 50 per cent. of the weight of the wood. In unseasoned ash and beech it is 20 to 30 per cent.; in loose-grained oak, hornbeam, maple, elm, Scotch fir, and spruce fir, 30 to 40 per cent.; in the looser fibred trees in which sap abounds, *e.g.*, the alder, the lime, the willow, and the aspen, 40 to 50 per cent.

The presence of water has generally a hurtful effect upon timber, as is shown in what follows.

B. The Changes which Wood undergoes.

The changes to which wood is subject are partly *mechanical* in their nature, consisting of alterations in the water capacity, and consequent alterations in shape; partly *chemical*, caused chiefly by the decomposition of the sap, which finally leads to the decay of the wood.

I. Changes in the Water Capacity, and the changes in form which are thereby produced.

Newly felled timber contains, as has been said, a large proportion of water—sometimes as much as 50 per cent. of its own weight. After lying for some time in a dry and airy place, it loses about half its amount of water by evaporation. Sawn or split wood, dried for a year or two under cover, still retains 10 to 15 per cent. of water, and only by continuous application of heat, or drying in an oven, can the water in timber be completely expelled.

Changes in the volume of timber.

During the process of drying, timber decreases in volume or *shrinks*. If exposed again to moisture it increases in volume or *swells*.

If any given piece of timber were uniform in texture throughout, and if no obstacles in any direction were presented to its expansion, the only result of shrinking or swelling would be alteration in volume; there would be no change in form. This, however, is seldom the case. Generally speaking, the texture of the wood varies in different parts of the same piece. Again, it is often used under conditions which do not permit it to shrink or swell freely in all directions; consequently, it shrinks or swells more in one place than in another.

When one part of a piece of timber shrinks more rapidly than an adjacent part, the wood *cracks*. If, on the other hand, one part swells more than another, or if the adjacent part meets with some obstacle to its expansion, the timber changes in shape—it becomes *warped*.

Shrinkage in different directions.

The shrinkage of timber stands in close connection with the amount of water contained. The more water it gives off while drying, the more it shrinks. Similarly the warmer and drier the air in which it is placed, the greater the shrinkage.

Some kinds of wood shrink more than others, and the same kind of wood shrinks differently in different directions.

All wood shrinks least in the direction of the fibres' length, and generally so very little that the difference need not be taken into consideration. But the difference caused by shrinking is very great across the fibres, and in *tangential section* it is two or three times greater than in *radial section*, or in the plane of the medullary rays. The sap-wood, which contains more water than the heart-wood, always shrinks more than the latter.

The following table, taken from "Karmarsch's Technology," shows the results of experiments made on a number of trees, to ascertain to what extent their timber shrinks. It must be observed that (1) the experiments were made with thin pieces of wood; (2) that the figures are understood to represent the difference between wood which is either quite green or saturated with water, and that which has been thoroughly well seasoned; and that, therefore, (3) the shrinking of partially seasoned wood *is considerably less* than is stated in the table. (The same applies of course to the swelling of such wood, when it is again exposed to moisture.)

The last column gives the average degree of shrinkage across the fibres.

Shrinkage of Timber.

Name of tree.	Degree of Shrinkage.			
	In length. Per cent.	Across the fibres in the direction of— The medullary rays. Per cent.	Across the fibres in the direction of— The annual layers. Per cent.	Average across the fibres. Per cent.
The common alder	0.369	2.91	5.07	3.99
The elm	0.124	2.94	6.22	4.58
The apple	0.109	3.00	7.39	5.19
The common ash (young)	0.821	4.05	6.56	5.30
The common birch	0.222	3.86	9.30	6.58
The common beech	0.200	5.03	8.06	6.54
The hornbeam	0.400	6.66	10.90	8.78
Ebony	0.010	2.13	4.07	3.10
The oak (young)	0.400	3.90	7.55	5.72
The oak (old)	0.130	3.13	7.78	5.45
The Scotch fir	0.120	3·04	5.72	4.38
The spruce fir	0.076	2.41	6.18	4.29
The lime	0.208	7.79	11.50	9.64
The common larch	0.075	2.17	6.32	4.24
The maple	0.072	3.35	6·59	4.97
Mahogany	0.110	1.09	1.79	1.44
Lignum vitæ	0.625	5.18	7.50	6.34
The pear	0.228	3.94	12.70	8.32
The rowan	0.190	2.11	8·88	5·49
The common walnut	0.223	3.53	6·25	4·89

General results afforded by the above table.

As is seen from the above table, the degree of shrinkage in the direction of the length of the wood is so slight that it may be left entirely out of consideration. In the direction of the breadth, however, it varies from 2 per cent. to 9 per cent. In radial section, the general average is 5 per cent.; for fir and pine 3 per cent.; for birch 4 per cent. In tangential section, where shrinkage is *greatest,* it varies from 2 per cent. to 13 per cent., the general average for wood in common use being 7 per cent.; for fir and pine 6 per cent.; for birch 9 per cent.

Fig. 4. Shrinkage in planks.

When a tree stem is sawn up into planks by parallel longitudinal cuts, the planks shrink as is shown in Fig. 4. The broadest portion shown, which includes the pith, shrinks least in breadth, most in thickness; least nearest the pith, most near the sides. The outermost plank, however, shrinks most in breadth—in the direction of the annual layers—and least in thickness. The planks lying between shrink differently on different sides, and become concave to the pith, and convex on the other side.

Of trees in most general use, beech, lime, hornbeam, and pear shrink *most;* birch, apple, white-beam, walnut, ash, and oak shrink considerably; alder, maple, Scotch fir, elm, spruce fir, and larch shrink in a *medium degree.* Mahogany shrinks least of all timbers.

Cracks occur in timber, because, as indicated above, it is seldom uniform in texture, and it is therefore liable to shrink in different degrees during seasoning. The parts nearest the sap-wood shrink more rapidly than the heart-wood, and cracks, which run almost invariably in the direction of the medullary rays, are the result. The more rapidly wood dries the more it cracks, consequently *timber should always be dried very slowly to prevent the formation of cracks.* If it is tolerably uniform in texture, it may, with proper treatment, be kept entirely free from cracks.

The *swelling,* or expansion of timber, takes place when it is exposed to damp air or water, and is in direct relation to its shrinkage. When a piece of dried wood is immersed in water, it swells until it occupies the same volume as it occupied in its fresh condition, after which no further expansion takes place. Its amount of water, however, and consequently its weight, are greater than in its fresh condition, because the vessels originally filled with air are now filled with water.

The *warping* of timber depends on differences in the nature of its texture, and on other circumstances which cause changes in form both when it shrinks and when it swells. For example, a plank will become twisted or curved if one side only is exposed to the sun without being turned. Thin, flat pieces of wood become convex or concave, according as one or other side is exposed to damp or to drying influences.

II. Means of preventing Cracks and Warping.

The means taken to keep timber as far as possible from cracking or warping during the process of seasoning, are very various. They are partly connected with the treatment of the wood when it is cut up into timber, and partly with its treatment for any special purpose.

1. Seasoning.

When wood should be cut down.

Trees should be felled when the sap is down or at rest. The best time is from the the middle of December to the end of February. Too much stress cannot be laid upon the importance of felling timber at the right time, for if felled at the wrong season, it will contain too much sap, which will make it very difficult to dry, render it much more liable to swell or shrink, and increase the risk of its becoming worm-eaten. In the case of needle-leaved trees excess of sap gives a bluish tinge to the surface of the timber.

Wood should be dried slowly.

The more slowly timber is dried the less it cracks, and timber felled at the proper season and allowed to *dry slowly* cracks very little. Barked timber, which dries more quickly than unbarked, often cracks so widely that it is quite unfit for slöjd-work. When the bark is left on, the cracks may be numerous, but they will be small. Thick pieces crack more than thin pieces; logs or round wood more than split wood;

sap-wood more than heart-wood. Care should be taken during seasoning *that the air has free access to the wood on all sides.* Wood which has been split with the axe is apt to crack at the ends; this may be prevented by pasting paper over them. Portions of timber containing the pith and the adjacent annual layers, always crack; such pieces are therefore unavailable for work. When round timber is split in order to facilitate seasoning, it should be divided through the pith.

Natural seasoning or seasoning by exposure to the air.

Boards or planks are best dried in a drying shed, where fresh air can circulate freely round each piece. The best way is to place the boards on their edges, with sufficient space between, taking care that they are not twisted in any way. If they are piled one on the other, pieces of dry wood should be placed between them, in order to separate them. For obvious reasons, none of the timber should touch the ground.

Timber which has been felled at the proper time, takes no harm from exposure to a little rain in spring and early summer, provided always that the air has free access, so that it may dry again quickly. Indeed, timber usually dries very rapidly out in the open air in early summer. The rain helps to wash out the sap, and the timber is thereby rendered more durable when thoroughly dried.

When wholly or partially finished planks are laid by for future use, care must be taken that they do not lie one close upon the other, but that both sides are fully exposed to the air, to facilitate further drying and prevent warping.

In the early stages of seasoning, evaporation goes on with tolerable rapidity, but afterwards it takes place more slowly, and timber must be kept in a dry and airy place for two or three years before it can be considered *fully* seasoned. Timber is said to be *seasoned* when the quantity of moisture it contains coincides with that contained in the atmosphere.

When timber can be said to be seasoned.

As has been said above, the amount of water in timber

seasoned as indicated, never falls below 10 per cent. of its weight. To decrease the water still further, it is necessary to dry the timber in ovens constructed for the purpose, or in heated air, or else to keep it for a long time in a warm place.

Influence of the sap on seasoning. Drying expels water only, not the essential elements of the sap, some of which part with great difficulty from water, and also take it up again with great readiness when the timber is once more exposed to moisture. These properties of the sap make seasoning much more difficult than it would otherwise be, and retard the process considerably in wood which abounds in sap—*e.g.*, beech, birch, oak, and walnut.

Removal of the sap. To overcome this difficulty, the sap may either be removed altogether, or its action may be neutralised. The first is accomplished by immersing the wood in cold water for some time, or in boiling water for a shorter time; or, what is still better, by *steaming* it. In the second case the timber is impregnated with substances calculated to counteract the destructive effects of the sap—*e.g.*, a solution of common salt, vitriol, chloride of zinc, etc. These methods can, however, only be mentioned here incidentally, as any detailed description would be entirely beyond the limits of this work.

2. Precautions necessary to prevent Warping and Cracking under special conditions.

As shrinkage is greater in tangential than in radial section, the wood for any special purpose ought to be sawn out or split in the direction of the radii of the stem, in order that the article may the better preserve its form and size. There are, however, some practical difficulties which render it impossible to carry out this principle in all cases.

Jointing pieces of wood. Uniformity of texture, and consequently less tendency to crack or warp, is more easily secured in small pieces of timber than in large pieces, and consequently it is usual in the construction of articles to employ smaller pieces of wood than are required, and to joint them together; and these pieces may often, without any disadvantage, be chosen from different kinds of wood, and may have their fibres running in different directions. Hence it is better in making a broad plane surface to select planks which have been divided in two, than to make it of whole planks. Planks containing the heart-wood nearest the pith which is generally cracked, are always divided in two to get rid of this portion.

Frames and panels.

Jointing also permits large plane surfaces to shrink without injury to parts of the work already completed. For example, blackboards, the panels of doors, etc., which are set into a groove in a frame, are thus permitted to shrink without cracking. Table-tops are strengthened by blocks which fit into a groove in the framing, and are glued to the under part of the top. Broad pieces of wood are furnished on one side with clamps, the fibres of which run at right angles to those of the broad piece, and which are inserted in such a way that the wood of the broad piece can shrink without hindrance.

III. The Decay of Timber.

After vital action ceases in a tree, its substance, like that of other organic bodies, undergoes a process of decomposition, which sooner or later terminates in the total *decay* of the wood. Decay takes place very rapidly if the timber is exposed to alternations of moisture, air, and heat.

The wood fibres themselves have a high degree of durability, especially if the sap, which is the prime cause of decay, has been removed. Some of the constituents of the sap, *e.g.*, starch and sugar, neither hasten decay nor retard it, while others, *e.g.*, tannic acid and resin, counteract it. It is the albuminoids which are the cause of decomposition, and the sap-wood in which they abound is the part which decays most rapidly.

Blue surface.

The *decay of timber* is caused, in the first instance, by the fermentation of the sap, which in this state soon acts injuriously on the wood-fibres. The first sign of this is a bluish tinge on the surface of the wood. Timber which has assumed this bluish tinge is not only less durable and strong, but it is also extremely difficult to work. Though the fermenting elements dry in the wood cells, they do not therefore lose their power. They remain dormant merely, and the application of moisture after the lapse of time is sufficient to wake them into activity. Hence, timber which is exposed to alternations of heat and moisture may very soon acquire a "blue surface," especially if kept where ventilation is deficient.

Dry rot. If the process of decay goes on further, fungi almost always make their appearance. One of the most destructive forms in which they appear is known as "dry rot."

Attacks of insects. Timber is also destroyed by insects or worms, which bore their way through the wood, and often reduce the inner portion completely to dust before any signs of destruction appear on the outside. Wood which is rich in sap, *e.g.*, birch and alder, is most liable to such attacks; beech is less liable; while the elm, the maple, and resinous needle-leaved trees, are seldom attacked.

Means of Preventing Decay.

As the decomposition of the sap is the real cause of the decay of wood, the means taken to prevent decay are directed either towards the retardation of this decomposition or to the complete expulsion of the sap, *e.g.*:

1.—The timber is cut down during the season of the year when there is least sap in the stem.

2.—The timber is seasoned as thoroughly as possible, in circumstances which permit free access and circulation of air, and is protected not only during seasoning but afterwards, from alternations of moisture and dryness.

The growth of fungus is prevented by exposure to light, and continuous and uniform ventilation.

3.—The wood, after it has been made into articles, is preserved from damp by varnish, oil paint, etc.

4.—The sap is got rid of by steeping the timber in water or steaming it in ovens.

It is to be observed, however, that in this way the constituents of the sap which *contribute to the durability* of the wood, *i.e.*, resin and tannic acid, are also removed.

5. The timber is impregnated with some substance in solution which neutralises the effects of the sap.

The two last named processes are not used for slöjd timber.

In conclusion, it may be added that when the sap is removed entirely, or when the timber is impregnated with some neutralising substance, it does not become worm-eaten. When insects attack wood which has not been treated in one of these ways, it is almost impossible to extirpate them. It has been recommended to apply an acid, *e.g.*. muriatic acid, or a solution of camphor to the worm-eaten holes; but this is, generally speaking, not practicable, and it is, moreover, not a complete cure.

C. Different kinds of Wood.

I. Comparison of the Qualities of different kinds of Wood.

The chief qualities of timber are:—*strength, the ease or difficulty with which it is split, hardness, toughness, elasticity, texture, colour and smell, weight, durability,* and its capacity for *shrinking* and *swelling.* The two last mentioned qualities have already been taken up.

It is obvious that most of these qualities depend not only on the kind of tree from which the timber is obtained, but also on many incidental circumstances, such as climate and soil, the age of the tree, the season of the year when it was cut down, subsequent treatment, etc. It is therefore hardly possible to make any general statements regarding them which shall hold good in all cases.

1. The **strength of timber** is shown by its power of resistance to *pressure, rupture, tearing,* and *twisting.*

The oak and the Scotch fir present the greatest resistance to *pressure.* The oak, the ash, the spruce fir, and next after them the Scotch fir, the larch, and the aspen, resist rupture best. In this respect the beech and the alder are not so strong. The oak and the ash, and after them the beech, the spruce fir, the Scotch fir, and the elm, present the greatest resistance to tearing.

2. **The ease or difficulty with which different kinds of wood may be split.** By this is meant the greater or lesser ease with which timber may be divided by a wedge-shaped tool in the direction of the length of the fibres. It is closely related to the quality of the fibres and the manner of their distribution. Wood which has grown quickly has long straight fibres, is free from knots, and is easily split. "Cross-grained" wood, the fibres of which twist and cross each other, and the wood of roots and of branches with knotty excrescences, is difficult to split. Wood from the lower part of the trunk nearest the roots is the most difficult of all to split.

When the medullary rays are large and long as in beech and oak, or numerous and fine as in needle-leaved trees, timber is easily split in radial section, but all timber is harder to split in tangential than in radial section.

The following timbers are *difficult to split* :—figured birch, hornbeam, elm, maple, and white-beam.

The following are *easy to split* :—ash, beech, alder, oak, aspen, Scotch fir, spruce fir, lime, poplar, and chestnut.

Old knotty oak, however, may present great difficulty.

3. The **density or hardness of timber** is shown in the resistance it offers to the tools with which it is worked. It is impossible to give definite statistics on this point, because

it depends so much on circumstances, *e.g.*, the varieties of texture in the same tree, the nature and arrangement of the fibres, the degree of moisture, the presence of resin, etc., etc.: the general rule, however, holds good, that close-grained timber with high specific gravity is hard (it being understood that comparisons are always made with seasoned wood). Seasoned timber is harder than green timber. Green heart-wood is harder than sap-wood. Resinous heart-wood is very hard, and this is also true of timber which has fine annual layers, as is shown especially in the extremely hard resinous knots often seen in planks.

Resistance to the axe and the saw.

The resistance which timber presents to the *axe* is greatest at right angles to the length of the fibres, and it decreases in proportion as the angle becomes more acute. It is least when the blade of the axe is parallel with the direction of the fibres' length, as in *splitting*.

The *saw*, on the other hand, works by *tearing* the fibres, and consequently it meets with most resistance in loose-textured timber with long tough fibres. Such timber makes the edge of the saw uneven. In close-grained timber with short fibres the saw works easily, and the edge keeps more even. Consequently, for heavy close-grained timber the saw does not require to be set so much. In certain kinds of timber moisture increases the toughness of the fibres, and on this account unseasoned timber is more difficult to saw than dry wood.

The *hardness* of timber is very important in all cases where it is exposed to blows, concussions, and general wear and tear.

For ordinary purposes the hardness of any piece of wood may be tested by cutting it with a knife.

The hardest timbers of all are lignum vitæ and ebony. The ordinary kinds of timber may be classified as follows:

Hard: hornbeam, maple, apple, pear, oak, and beech.

Medium: ash, elm, white-beam, walnut, birch, lime, and chestnut.

Soft: Scotch fir, spruce fir, larch, alder, aspen, and poplar.

As has, however, been indicated above, spruce fir with fine annual layers and resinous Scotch fir are often very hard, and they might thus find a place in the higher class.

4. **The toughness and elasticity of timber.** A piece of timber which may be bent without breaking, and which does not resume its former shape when the bending force is removed, is said to be *tough;* if it does resume its former shape, it is said to be *elastic.* Generally speaking, both these qualities co-exist in all timber, but one is usually more predominant than the other, according to the kind of wood. Thus some timbers are said to be elastic and others tough.

Unseasoned wood is tougher than dry wood, and what it gains in elasticity during seasoning it loses in toughness. Damp heat increases toughness; hence hoops and sticks are "steamed" in order that they may be bent.

As a general rule light timber is tougher than heavy timber, roots are tougher than stems; sap-wood is tougher than heart-wood, and young timber is tougher than old.

The toughest timbers are the following:—hornbeam, elm, ash, aspen, birch, juniper, hazel, osier, maple, and white-beam.

Lime, alder, beech, and the heart-wood of oak are only moderately tough.

Elasticity is increased by seasoning, and is generally great in heavy timbers. It is of great importance in the manufacture of many articles, *e.g.*, masts, oars, wooden springs, the handles of spades, axes, hammers, etc.

The following timbers are *elastic:* elm, ash, aspen, oak, spruce fir, birch, maple, and poplar.

Hornbeam, alder, and Scotch fir are less elastic.

5. The **texture, colour** and **smell** of **timber.** Knowledge of these qualities is very important in connection with the recognition of different kinds of timber, and in estimating their value

By *texture* is understood the way in which the vessels, fibres, medullary rays and annual layers are *woven* or connected together. (See fig. 1).

Wood as it appears in cross section is said to be **end way of the grain**; as it appears in radial and tangential section—parallel with the fibres—it is said to be **length way of the grain, or with the grain**; and as it appears when we look across the fibres at right angles to their length, it is said to be **across the grain**.

Coarse and fine texture

We distinguish between *coarse* and *fine* texture according to the *quality* of the fibres, vessels, medullary rays and annual layers, which, taken all together, give to wood its characteristic appearance. Similarly we speak of *long-fibred* and of *short-fibred* texture, according as the wood "works" with long or short shavings.

Different colours of wood.

The colour of wood varies from *white* to deep *black*, with many intermediate shades of yellow, red, brown, etc., depending on the kind of tree. It varies not only in different kinds of timber, but in the same kind of timber, and even in the same tree. As has been said above, the heart-wood is always darker than the sap-wood. Certain kinds of timber, again, *e.g.*, oak and mahogany, become darker with time.

Our ordinary timbers are whitish, yellowish, brownish or reddish, and are not so highly coloured as tropical timbers, some of which are very striking in colour.

The smell of wood due to the sap

The *smell* peculiar to many kinds of timber is a mark by which they may sometimes be recognised. This characteristic smell does not proceed from the wood itself, for it has none. It is due to the sap, and is always strongest in fresh sappy wood; though seasoned timber sometimes has a very decided smell, which is often quite unlike that of the unseasoned wood. Needle-leaved trees have a strong smell of turpentine, and certain broad-leaved trees, *e.g.*, the oak, often smell of tannic acid. Many trees have an agreeable smell, *e.g.*, the cedar, juniper, the camphor-tree, etc. The smell of some timber remains in it for a long time, and communicates itself to food kept in vessels made of it.

A *musty smell* in timber is a sign of decay.

6. **The weight or specific gravity of timber** is very variable, depending as it does on a number of different circumstances. Hence it is impossible to give such definite statistics under this head as can be given in the case of metals and many other substances. We have to take into consideration the closeness or the looseness of the fibres, which determines the hardness or density of the wood; the presence of more or less sap; the climate and soil in which the tree has grown; its age; its different parts; the degree of seasoning, etc.

Specific gravity of the cellular tissue.

The specific gravity of *wood* properly so called, *i.e.*, of the cellular tissue which composes it, is very similar in all timbers, and even in the lightest kinds it is greater than that of water. Nevertheless, most timbers, owing to their porous nature, are lighter than water, and float in it. This is the case with all our indigenous trees after seasoning. A warm climate produces heavy timber; and the heaviest timbers, such as ebony and lignum vitæ, are found in the tropics.

The presence of water is the circumstance which most affects the weight of timber. All timbers are heavier when newly felled than after seasoning. Hence, in determining the specific gravity of different kinds of timber, we must assume that the timber is fully seasoned.

The *average* specific gravity of the most common kinds of timber is given as follows by competent authorities:—

	NEWLY FELLED.	SEASONED.		NEWLY FELLED.	SEASONED.
The Hornbeam ...	1.08	0.72	The Spruce Fir ...	0.73	0.47
The Common Alder	0.82	0.53	The Lime	0.74	0.45
The Elm	0.95	0.69	The Common Larch	0.76	0.62
The Apple ...	1.10	0.75	The Maple	0.90	0.66
The Common Ash	0.92	0.75	The White-beam ...	1.04	0.86
The Aspen ...	0.80	0.49	The Pear	1.01	0.72
The Birch ...	0.94	0.64	The Rowan... ...	0.96	0.67
The Common Beech	1.01	0.74	The Common Walnut	0.91	0.68
The Oak	1.10	0.86	Ebony	—	1.20
The Com. Juniper	1.07	0.61	Mahogany	—	0.81
The Scotch Fir ...	0.70	0.52	Lignum vitæ ...	—	1.40

The *absolute weight* per cubic foot in any given timber is ascertained by multiplying the specific gravity given above by 62.5 = the number of pounds in a cubic foot of water.

7. **The durability of timber.** This and the circumstances which favour it have been touched on in connection with the sap, with seasoning, with decay, and the means of its prevention.

The conditions under which timber is used have the greatest influence on its durability. Thus, timber which is kept *under cover* and protected from moisture is very durable, and may last for many centuries. Some kinds of timber are extremely durable if kept under water. Thus, the oak used in ancient lake-dwellings and bridges, or found in bogs, has been preserved for thousands of years. *The durability of wood under water.*

If timber is exposed to *alternations of moisture and dryness*, its durability is diminished; and yet, in most cases, it is precisely in these unfavourable conditions it has to be used. *When wood is least durable.*

Hence it follows that it is impossible to give precise details regarding the durability of timber. Under this head all that can be done is to mention the trees which in all circumstances give the most durable timbers, viz.: the *oak*, and resinous, close-grained *Scotch fir* and *larch*. The *elm* comes next to these. If exposed to alternations of moisture and dryness, oak is said to last one hundred years, birch fifteen years, and beech not more than ten. Durability is also mentioned in the description of different kinds of timber, which follows. *The most durable timbers.*

II. Characteristics of different kinds of trees.

Here follows an enumeration of the different kinds of wood which are available for slöjd work, together with a condensed statement of their properties, in order that, as far as is possible in a brief description, the reader may be made acquainted with each kind of timber.

[The following kinds of wood can be easily obtained in

England, and are therefore specially recommended :—Scotch fir, spruce fir, alder, birch, beech, oak, chestnut, lime, and poplar. See also p. 204.—TRS.]

1. Needle-leaved Trees.

The **Scotch fir** *(Pinus sylvestris).*—The ripe timber is yellowish white or reddish white. The boundaries of the annual concentric layers are light brown in the heart-wood; white in the sap-wood. It is the heaviest, hardest, and most resinous of all the needle-leaved trees, and has a tolerably strong smell of turpentine. Its resinous, fine-grained heart-wood is very durable.

The **spruce fir** *(Pinus abies).*—The wood is yellowish white. In a longitudinal section it shows dark reddish streaks. It is very elastic, and is easily split with the axe. As it contains a good deal of resin, it resists damp; though, being less resinous than the pine, it is more easily glued. Like the pine, it makes excellent timber. Very hard knots, which loosen and fall out when the wood is seasoned, are, however, of frequent occurrence in this wood.

The **common larch** *(Pinus larix).*—The wood of this tree is reddish, with dark annual layers and white sap-wood. It warps but little, and does not readily become worm-eaten. It is more durable than the Scotch fir and the spruce fir.

The **common juniper** *(Juniperus communis).*—The wood of the young bushes is white, and it deepens from yellow to brown as it increases in age. It is hard, tough, close, strong, and durable, and whenever it can be obtained large enough it is much in request for slöjd articles. The juniper has a peculiar and agreeable smell.

2. Broad-Leaved Trees.

The **hornbeam** *(Carpinus betulus).*—The wood of this tree is white, very hard, heavy, close and very tough. The medullary rays are very little darker than the wood, and are not easily distinguished. They are curved, appearing in a longitudinal section like narrow inconspicuous flecks. The wood is very difficult to split. It dries slowly and warps easily. It is very durable if kept dry, and is a favourite timber for slöjd work.

The **common alder** *(Alnus glutinosa)*.—The wood is whitish or brownish-yellow, often deepening to brown, and in the newly-felled tree light red. The annual layers are difficult to recognise; the medullary rays are rather broad, and brown in colour. The timber is only of medium hardness, and is neither very tough nor very elastic; it splits readily, and does not crack or warp easily. It is very durable if constantly kept wet, but it is of low durability if exposed to alternations in the degree of moisture. If felled at the wrong time it is speedily attacked by worms. Its close and even texture make it good timber for slöjd work.

The **hoary-leaved alder** *(Alnus incana)* furnishes timber which is whiter, finer, and closer than the preceding.

The **elm** *(Ulmus montana, U. campestris)*.—The colour of the young wood in general, and of the sap-wood in older trees, is whitish-yellow. The old heart-wood is reddish-brown, streaked and veined. The inner boundary of the annual layers is somewhat lighter in colour and looser in texture than the rest, and has visible pores. The medullary rays are very narrow and numerous, giving to this timber in longitudinal section a dotted and streaked appearance. This timber is moderately fine in fibre, tough, hard, given to warp, difficult to split, and not liable to the attacks of worms. Its durability under all circumstances is very great. It is often beautifully marked.

The **common ash** *(Fraxinus excelsior)*.—The colour of the young wood is white; of the older, yellowish brown, deepening almost to brown in the heart-wood. The medullary rays are not easily distinguished. The annual layers are generally broad, and, as in the case of the oak, the large pores on their inner edge render them very conspicuous. This timber is tough, elastic, very hard, easily split, not liable to crack, and, if kept in a dry atmosphere, extremely durable. If exposed to the open air it is of low durability. It is much esteemed for its strength and toughness, and is used with advantage for springs of all kinds, tool handles, etc., etc. The young wood is used for barrel-hoops, etc.

The **aspen** *(Populus tremula)*.—The wood is white, with coarse annual rings. It is fine in texture; tough, easily split, and warps but little. It is very durable if kept under cover or in the ground. It is not of much use in slöjd work, and in Sweden it is used chiefly in the manufacture of matches.

[The **poplar** *(Populus)*.—The colour of the wood is a yellow or brownish white. The annual rings are a little darker on one side than on the other, and are therefore distinct. The texture is uniform, and there are no large medullary rays. The wood is light, soft, easily worked, and does not splinter. When kept dry it is tolerably durable, and it is not liable to shrink.—TRS.]

The **common birch** *(Betula alba)*.—The wood of the young tree is white. Older wood is reddish white in colour. The medullary rays are very narrow and scarcely distinguishable. The timber is tolerably hard, and very tough; it dries very slowly, and swells easily. It is very durable if kept dry, but is of low durability if exposed to the open air, and is very apt to become worm-eaten.

The quality of birch varies very much, and depends greatly on climate and soil. Birch grown in favourable soil is straight in fibre, easily split and easily worked:—Birch grown in dry and stony ground or in marshy places is crooked in fibre and more or less knotty, gnarled and cross-grained, and difficult to split. Timber of this kind is beautifully marked. In most parts of Sweden birch furnishes the greater proportion of the wood used in slöjd, and takes the place of the beech and the hornbeam of southern Sweden and southern countries.

The **common beech** *(Fagus sylvatica).*—The wood in the young tree is light brown; old wood is very dark. The medullary rays are large, glossy, and dark brown, and the general colour of the wood is uniform. The concentric annual layers are not specially conspicuous, but they are easily distinguished. Beech timber is hard, close, heavy, and easily split, especially in the direction of the medullary rays. It is inelastic and rather brittle. It dries *very* slowly, and warps easily. It is very durable under water and when kept dry, but if exposed to varying degrees of moisture it is the least durable of all timbers. It is highly valued for its hardness, and much used for barrels.

The **oak** *(Quercus robur).*—The sap-wood and the wood in young stems is nearly white. The heart-wood in older trees is brownish. The large pores on the inner edges of the annual layers, and the broad, yellowish brown, frequently glossy, medullary rays are specially noticeable. This timber is peculiarly hard, strong, and durable. It is not affected by alternations in the degree of moisture, and it is in all circumstances the *most durable* of all our timbers. It dries slowly, and is very apt to warp unless thoroughly well seasoned. After being in water—especially salt water—for many years, its colour becomes bluish black. The oak furnishes better timber than any other tree of Northern Europe.

[The **chestnut** *(Castanea vesca).*—The colour of the sap-wood is yellowish white; that of the heart-wood is light to dark brown. The wood of the chestnut resembles that of the oak in colour, but it may easily be distinguished from it by the absence of the broad medullary rays which are found in the oak. The timber is heavy, hard, elastic, and very durable if kept uniformly either dry or wet. If subjected to variations in the degree of moisture it is of low durability. —TRS.]

The **lime** *(Tilia)*.—The wood is usually white, soft, and light. The medullary rays are extremely fine, and the annual layers can scarcely be distinguished. It does not warp easily. It is of low durability, and is not very serviceable.

The **maple** *(Acer platanoides)*.—The wood is white, with very narrow and numerous medullary rays of a faint brown colour, which give it a beautifully "waved" lustrous appearance. The annual layers are inconspicuous. The wood is uniform in texture, hard, strong, tough, and difficult to split; it presents a glossy surface to the plane, and does not crack or warp readily. In consequence of these good qualities, it is much sought after for slöjd timber.

The **white-beam** *(Sorbus Scandica)*.—The wood of the young tree is yellowish. Older wood is light brown or reddish in colour. It is frequently speckled or veined. This timber is fine and uniform in texture, hard, close, and very tough. It warps but little, and is much valued as slöjd timber.

The **pear and the apple** *(Pyrus)*.—The wood of the young tree is nearly white. Older wood is dark brown, sometimes red in colour, and often streaked. It is very fine and close in texture, hard, heavy and tough. The medullary rays are small, and they and the annual layers are inconspicuous. It can be cut easily in all directions, and does not splinter, owing to the uniformity of its texture.

The wood of the apple tree has a general resemblance to that of the pear, but it is closer, redder, and harder—indeed the apple furnishes one of the hardest timbers. The wood of the wild pear or apple is superior to that of the cultivated varieties. The wood of both trees is much esteemed.

The **rowan** *(Sorbus Aucuparia)*.—The wood is whitish or light brown. In some respects it resembles the white-beam, but it is not so good. As slöjd timber it may often rank with the birch.

The **common walnut** *(Juglans regia)*.—The wood of

the young tree is almost white, loose in texture, and soft. Older wood is brownish grey or dark brown, and is often beautifully marked. It is hard and strong, and generally close in texture, though, like the oak, it has particularly large pores. The medullary rays are almost invisible. It dries very slowly, and shrinks a good deal. It is one of the most beautiful European timbers, and is extensively used.

The following tropical timbers may also be mentioned:—

Ebony *(Diospyros)*.—From Africa and the East Indies. The sap-wood is quite white, the heart-wood generally quite black, though sometimes brownish black with white streaks and flecks towards its inner edge, which detract from the value of the wood. Its texture is so uniform that it is impossible to distinguish the annual layers or the medullary rays. The timber is brittle, but very hard, close and heavy. On account of the three last named qualities, and its beauty, it is much esteemed, but it is too expensive to be used to any great extent.

Mahogany *(Swietenia Mahogani)*.—From Central America and the West Indies. Other kinds of timber are also sold under this name. When fresh the wood is generally reddish or brownish yellow, but it gradually darkens, and finally becomes almost black. It has narrow, rather inconspicuous annual layers, and small but distinctly visible pores. In longitudinal section the figuring of this timber is very beautiful. It has fleck-like or pyramidal markings, with a fine satin-like lustre. It varies much in hardness, weight, closeness, and general texture in different varieties. Mahogany is under all circumstances very durable. It warps but little; shrinks *less* than any other timber; and is never attacked by worms. It is highly esteemed as timber, and is very extensively used.

Lignum vitæ *(Guaiacum officinale)*.—From Central America. The wood is greenish or blackish brown, with yellowish and dark streaks in longitudinal section. It is heavy, resinous, very close-grained, and almost as hard as metal. It is twisted in fibre, very difficult to split, and therefore not easy to work. Its extraordinary hardness and great durability make it valuable in the case of articles which are exposed to much wear and tear.

CHAPTER III.

TOOLS.

A. Choice of Tools.

Choice of tools for slöjd teaching.

The tools used in slöjd teaching must be chosen with due regard to the pupil's capacity. They ought to be neither too large nor too heavy, but such as can be easily handled. It might perhaps be considered advisable to use tools slighter in make than those generally employed in slöjd-carpentry, and the question might be raised whether such small tools as are to be found in "children's tool-boxes" should not be procured. Tools of this description are, however, usually too inferior to be taken into consideration at all; and, if specially ordered in a good quality, they would be much dearer than those sold in the ordinary course of trade. This applies particularly to tools made of iron or steel. Moreover, such small tools are particularly difficult to keep in order, because they are very slight and brittle. And further, a little experience in teaching proves that children from eleven to fourteen years of age require tools quite as substantial and durable as their elders. Whether or not a tool is too heavy depends upon the person who uses it, for one child may have the strength required to use a much heavier tool than can be used by another. In connection with this it should be noted, that if children are not accustomed, while receiving instruction, to use and to keep in order the tools used in ordinary life, it will be very difficult for them to manage them when they are older. It may be objected that if children use the ordinary knife, saw, axe, etc., they may easily hurt themselves; but this is quite as likely to happen with "toy tools." Besides, it is the duty of the teacher to insist that the children pay attention to the manner of using the tools, and use them in such a way that they do not hurt themselves.

Size of tools. Although we maintain that the tools used in slöjd teaching should be of the size generally employed, it does not therefore follow that the *largest size* is to be selected, but rather that the *smallest* should be chosen, such as the little hands of the youthful pupil can efficiently wield without much trouble. The handle of the knife should not be larger than can be grasped, though the blade may be of the usual size. The smoothing plane should be 7½ inches long and 2¾ inches broad. The trying plane should not be unnecessarily long; 22 inches is long enough, though the breadth ought to be 3½ inches, or broad enough for an iron of 2½ inches. If the trying plane is narrower, it is difficult to plane a surface of any size, and the smaller tool would occasion more work and trouble than one of the dimensions given above. The handles of chisels and similar tools should not be larger than is necessary. The axe should not weigh more than 2 lbs. The frames of the bow saws should be of the lighter description of those used in carpentry.

As one of the aims of slöjd teaching is to develop the physical powers of the pupil, each separate exercise must lead up to the next in such a way that the pupil proceeds from easier to more difficult work. But the most perfect gradation of exercises arranged on this principle will not ensure success if the teacher does not know how to choose suitable wood for the pupils' work, and does not take care that they have *good tools* in good condition. As we demand of the pupils work well executed and accurate in all its details, we are bound to see that they are provided with suitable wood and good tools.

As regards suitable wood, the reader is referred to Chapter II. It need be merely named here that the wood must be *sound, well seasoned, straight in fibre,* and, as far as possible, *free from knots.*

The tools selected should always be of the best quality, even if these should prove rather more expensive. Instead of buying a large number of inferior tools at once, a few good ones should be procured. But it is not enough to *buy* good tools, they must be *kept in good order.* Ability to keep tools in order is an *indispensable qualification* in a good teacher of slöjd, for if he lacks skill in this respect his teaching will also lack one of the first conditions of success. There are two rather complicated tools which are particularly difficult to keep in order, *i.e.*, the *plane* and the *saw.* A great deal of energy is wasted in slöjd teaching if the pupils work with badly set planes or with blunt saws. Hence special care should be bestowed on these tools.

Quality of tools.

Good tools indispensable.

Practice in grinding tools and keeping them in order must be included in the instruction given. Great demands in this respect must not be made at first, but they may be gradually increased until the pupils, at least towards the end of the course, are able to grind a plane iron and sharpen a saw. If this is expected of the pupils, so much the more must it be demanded of the teacher.

Grinding tools.

The description which follows attempts to give, to some extent, detailed knowledge of the tools which are used in educational wood-slöjd, together with instructions for keeping them in good condition. The illustrations accompanying the description are taken from selected tools and appliances, and the scale is indicated by the fraction after the name of the figure. Want of space prevents the insertion of complete representations of all the tools, etc. A few illustrations of this kind, particularly of benches and of a cupboard for tools, have been added on separate plates at the end of the book, for the guidance of those who wish to make these articles. The technical names are, generally speaking, those employed in carpentry; but a proportion of the names of tools, exercises, and methods of manipulation, have originated and been adopted in the course of the development of slöjd teaching.

B. Appliances for holding the work.

1. **The bench** is the article most frequently used for holding the work steady during its execution. It is the most indispensable part of the apparatus required for slöjd.

Fig. 5. Bench. 1/20.

A bench top, *B* front bench vice, *C* back bench vice, *D* bench well, *E* bench drawer or till, *F* front rail of bench box, *aa* bench pegs or hooks, *bb* holes for bench pegs, *c* vice tongue or key, *ee* screw-bolts, *f* back rail of bench box, *gg* vice-screws, *h* front rail of bench.

The *Single Bench* (Fig. 5) is practically a strongly constructed table, heavy enough to stand steady during the work.

A complete bench.

The bench top consists of a strong, hard, close piece of plank about 3 inches thick. For the purpose of holding the work fast it is provided either with one screw or two, arranged in a particular way, called the *back bench vice* and the *front bench vice.* A complete bench (Fig. 5) has both; one of simpler construction (Fig. 8) has only the back bench vice.

Back bench vice.

At one end of the bench-top, to the right of the worker, a rectangular piece is cut away from the anterior edge, its length being parallel to the edge, and where this piece has been cut away a prismatical frame-work is moved by the turning of a wood-screw. The nut into which this screw catches is firmly fixed to the end of the bench top. The frame-work is directed partly by the screw, partly by separate *bolts*, and the screw is held fast by means of a *wedge* or flat pin, which catches like a fork in a groove on the screw.

This arrangement is called the *back bench vice.* The frame-work is perforated perpendicularly by one or more square *holes*, from 4 to 6 inches apart, and a row of similar holes is introduced in the bench top, in a straight line with those in the frame-work. When a plank is to be held in a horizontal position on the bench, a *bench peg* is placed in a hole in the bench vice, and another in a hole in the bench top at a distance corresponding to the length of the plank, and the screw is applied. Care must be taken that the head of the bench peg does not rise above the upper surface of the wood, and also that, during planing, the iron of the plane does not come in contact with the head of the peg, a fault often committed through carelessness by beginners.

The bench pegs (Fig. 6) are rectangular pieces of iron from 8 to 10 inches long, which fit rather loosely into the holes of the bench top, and are provided on one side with a steel spring, in order that they may remain fixed at any desired height. The head of the peg is double-grooved, to hold the work securely. To make room for the head of the peg, the holes in the bench top are usually sufficiently enlarged at the upper end to permit the head to be pushed down, until its top is level with the bench top.

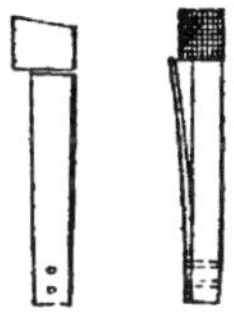

Fig. 6.
Bench Peg. 1/10.

Front bench vice.

The arrangement of the screw, to the left of the worker, is termed the *front bench vice.* It is much simpler in construction than the *back bench vice.* Fig. 5 shows its construction. A movable piece of wood is placed in front of the end of the screw, called the *vice tongue* or *key* (Fig. 7), partly to hold the work more securely, partly to prevent its being injured by the screw. When a long piece of wood is fastened into the front bench vice for edge-planing, it is advisable to allow the under edge

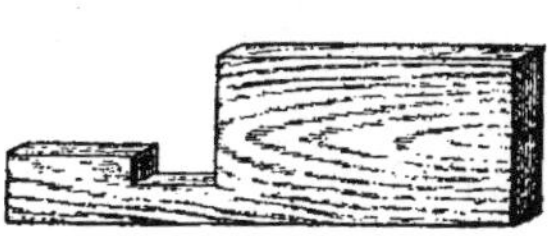

Fig. 7. Vice Tongue or Key. 1/10.

to rest upon a little block on a swivel, attached to the under side of the bench top. If the screws do not turn easily, the friction may be reduced by rubbing them well with pulverised plumbago.

On the side of the bench farthest from the worker is a trough or channel, called the *bench well,* in which tools not in actual use may be laid. Triangular pieces of wood, firmly attached to the ends of this well, facilitate the sweeping out of shavings, etc.

The different portions of the bench are fastened together by dovetailing, mortising, and iron screws.

The bench top rests upon feet or *rails,* and it is often furnished on the under side with a drawer or till. A similar drawer may be connected with the rails.

The wood used for the bench top should be oak, ash, beech, or hard pine; for the screws, horn-beam or "figured" birch; for the well and the rails, fir or pine.

The complete bench described above is too large for general use in school slöjd, the space for which is usually limited. As only one person can advantageously work at it, it is also too expensive.

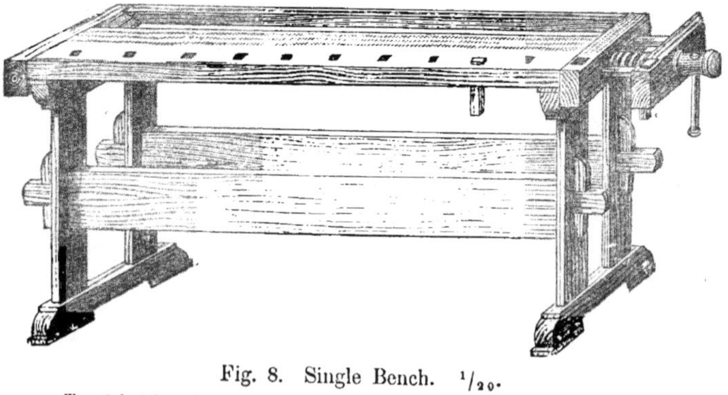

Fig. 8. Single Bench. $^1/_{20}$.
Top, 5 feet long by 1½ feet broad. Height, 2 feet 7 inches. Nääs pattern.

The bench represented in Fig. 8 is more suitable for schools

Bench after the Nääs pattern.

where many benches are required. It is at once simple and practical. It takes up little space, and it can be procured for one-half—indeed for one-fourth—of the cost of the bench first described. It is furnished with a back bench vice only, consisting of a piece of wood moving on bolts, and worked by a screw fixed with a forked *wedge* to the movable front jaw of the vice. The bolts must be firmly inserted in the detached portion of the vice, and must have their anterior ends made fast in a cross-piece; otherwise the movable portion of the vice will not move easily and surely backwards and forwards by means of the screw. To fasten a piece of wood quite steadily in the vice it should be balanced as nearly as possible on the top of the screw. When this is not done, it has a tendency to fall to one side, and if this frequently happens the vice will finally be destroyed.

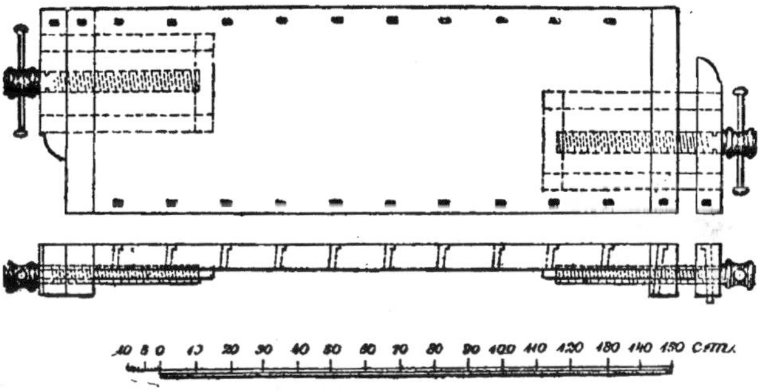

Fig. 9. Double Bench.

Adjustable double bench.

This bench may also be adapted for two persons by introducing a screw in each end of the bench top, as indicated in Fig. 9. The bench top in this instance ought to be rather broader than in the preceding. The height of the bench ought to be adapted to the height of the worker, and accordingly separate pieces of wood, provided with hinges, are attached to the upper or lower cross-bars of the feet, and by the raising or letting down of these the bench top is raised or lowered.

R. Trainor's bench.

Fig. 10 is a bench of English manufacture, well adapted for slöjd work, and is known as R. Trainor's Improved Bench.

Fig. 10. Trainor's Bench *.

A bench top, *B* tool tray or bench well, *C* back strip, *d* tail (or back) bench vice, *e* side (or front) bench vice, *f* plane rest or fillet, *g* Merrill's bench stop, *hh* bench pegs, *ii* joint bolts, *MM* fore legs, *NN* rear legs, *O* front bottom rail, *P* back bottom rail.

This bench is constructed so as to be portable. It consists of a hard wood top *A*, 4½ inches thick, made of beech or birch, and is supported by a strong framework *MM NN PO* made of fir, and bolted and framed together. The forelegs *MM* are "strutted," in order to prevent the framework from shaking loose through constant use and pressure. The bench is 5 ft. long, and 2 ft. wide; and it can be made from 2 ft. 6 in. to 3 ft. high.

The side or front bench vice, *e*, attached to the bench is made of metal, and is called "Crossley and Macgregor's Patent Instantaneous Grip Vice." The tail or back bench vice, *d*, is of German pattern, and acts as a cramp vice in conjunction with the bench pegs *hh*. Only the screw part of this vice is made of metal.

As was said above, this bench is well adapted for slöjd work. It stands firmly in position without being screwed to

* For prices of this bench see p. 215.

the floor; its vices, pegs, and stops are all new designs, and being made of metal, they are easy to work, and do not readily get out of order. The space *Z* underneath is specially constructed to admit of the fitting up of lookers and drawers.

The **Holdfast** is a simple appliance which is often used to secure pieces of wood to the bench in sawing, boring, chiselling, etc. The holdfast (Fig. 11), consists of a round iron or steel rod, furnished at the upper end with a strong arm. It is inserted in a hole bored in the bench top, the diameter of which is very little larger than that of the cylindrical portion of the holdfast. The piece of work is laid under the arm, and secured by a stroke from the mallet on the heel, in the direction *a*, and is loosened by a stroke in the direction *b*. The holdfast may therefore serve the same purpose as the back bench vice.

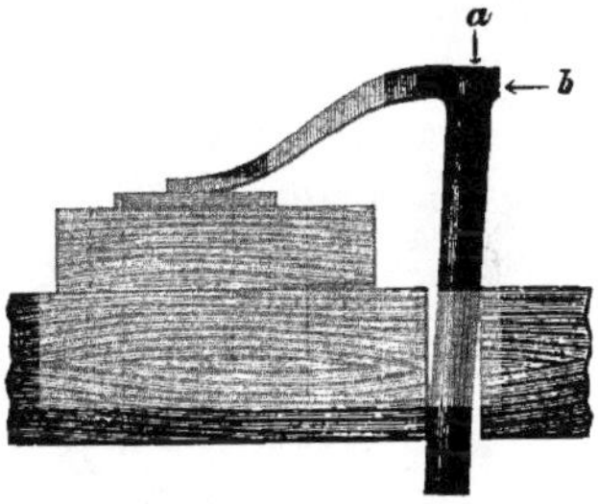

Fig. 11. Holdfast. 1/8

The **shooting board** is a contrivance which may be advantageously used when a partially planed piece of wood has to be squared up at right angles to a plane surface or a straight edge.

The shooting-board (Fig. 12) consists of a piece of hard pine $1\frac{1}{2}$ inches thick, 8 inches broad, and from 2 to $2\frac{1}{2}$ feet long, on one side of which there is a rebate, which serves as a guide to the trying plane when in action. At the further end there is a smooth rectangular block, the inner side of which is carefully secured at right angles to the rebate of the plane rest. Under this plane rest a groove is hollowed out, in order that the shavings may not pro-

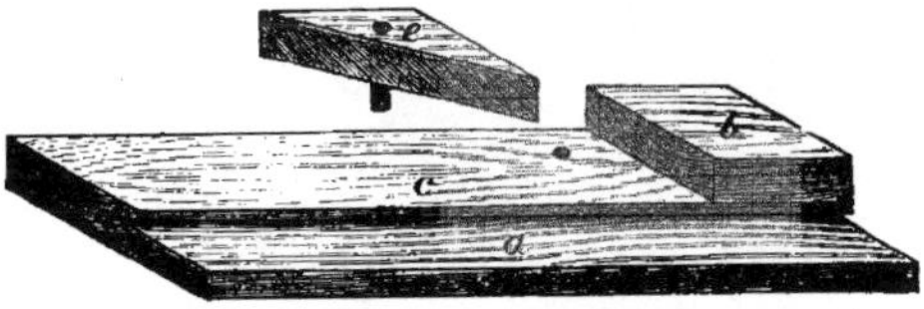

Fig. 12. Shooting-board, 1/15.

a. Plane rest, *b.* Block for square shooting, *c.* Rest for wood, *d.* Rebate and groove, *e.* Block for mitre shooting at an angle of 45°.

vent the plane from lying close to the rebate during work. Instead of a rebate made in a thick piece of wood, two pieces may be fastened together, a narrower above a broader piece. In this way a rebate will be formed. Before they are fastened together, the under part of the inner edge of the top piece must be cut away so as to form the groove for shavings.

When the shooting-board is in use, it is secured between two bench pegs. The piece of wood which is to be squared is held and pressed against the trying-plane with the left hand, the plane being directed by the right. Care must be taken not to plane anything off the edge of the rebate, or to hurt the fingers.

The shooting-board may also be used for mitre shooting pieces of wood which are to be fastened together at an angle of 45°, by placing before the block for square-shooting a triangular block whose anterior edge forms an angle of 45° with the edge of the rebate. See Fig. 12.

II. Handscrews.

Handscrews are used to secure the work to the bench, and to hold several pieces of work fast while a drawing is being made or while glue is drying. The bench itself, when not otherwise engaged, may be used with advantage in the case last named.

Handscrews are made of wood or of iron, and are of various sizes.

Wooden handscrews (Figs. 13, 14), consist of three straight pieces of wood, two of which are joined to the third on the same side, and at right angles to it. Horn-beam or tough birch is the best wood for the purpose. A strong wooden screw passes through one of the parallel arms and gives the necessary pressure.

As the handscrew is sometimes subjected to a greater strain than the construction just described can bear, it is often strengthened by an iron rod. (See Fig. 14.)

Fig. 13. Handscrew. $\frac{1}{8}$.

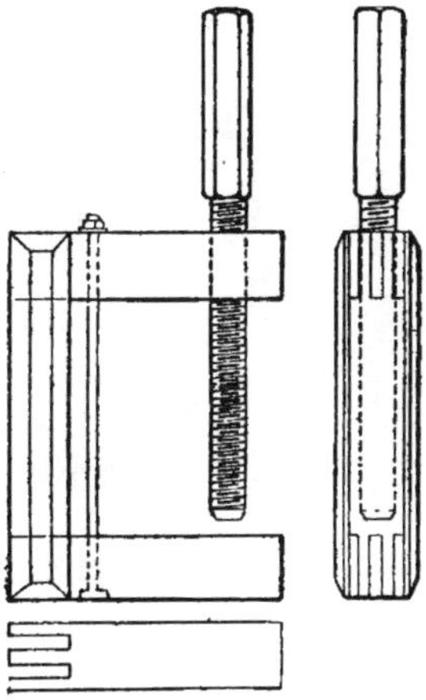

Fig. 14. Handscrew. $\frac{1}{8}$.

Fig. 15. Adjustable Handscrew. $\frac{1}{16}$.

When the screw is applied, one hand only should grasp the handle, and the other should take hold of the screw either above or below the nut. Otherwise, if the pressure is great, the screw may break. If the screw should go off the straight during the process, a light blow from the mallet on the lower part will put it right. A piece of wood should always be laid under the point of the screw, to prevent marks on the work.

[The English handscrew (Fig. 16) differs from the Swedish handscrew in having two screws *a a* instead of one. These screws work in opposite directions, through two square wooden cheeks, *b b*.—Trs.]

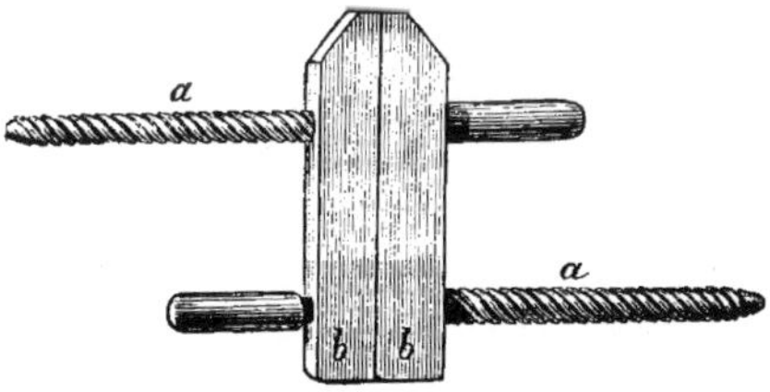

Fig. 16. English Handscrew.
a a screws, *b b* cheeks.

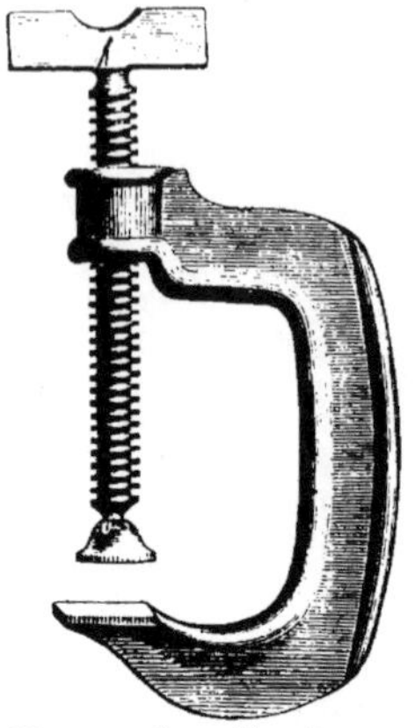

Fig. 17. Iron Handscrew, or Thumbscrew cramp. $\frac{1}{3}$.

Thumbscrew cramps are now made of wrought iron. This gives strength without weight or clumsiness. These screws are very useful, and easily managed. (See Fig. 17).

When broad pieces of wood have to be glued together, and the handscrews already described are not large enough, and the bench is not available, use is made of a screw in which a movable block is substituted for one of the parallel arms. Such screws are called **adjustable handscrews** (see Fig. 15).

C. Setting out.

It is often necessary for accurate workmanship to draw or mark the outlines of the pattern object on the wood, at various stages of the work. This is done by tracing round the outline of the model, by copying a drawing, or by means of given measurements.

The following tools are necessary :—

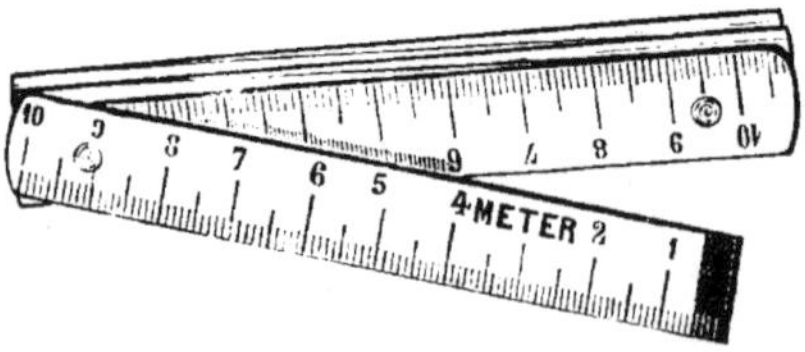

Fig. 18. Folding Metre-measure. $\frac{1}{2}$.

I. The **metre-measure** for measuring off and subdividing measurements. A rule of hard wood, one metre or half a metre long, divided into centimetres and millemetres, is the best for the purpose. A thin folding rule of strong wood or ebonite may be used for less exact measurements, and is convenient to carry about, but is not altogether trustworthy, on account of the looseness of its construction, and the gaps at the joints.*

Fig. 19. Marking-point 1/3.

II. **In drawing straight lines** use is made of an ordinary **ruler** and a lead pencil, but when great accuracy is required a **marking point** should be used. This consists of a piece of steel, tapering to a sharp point, about 4 inches long and ¼ inch thick, inserted in a handle (Fig. 19).

III. In drawing lines parallel to the edges of a piece of wood, the **marking gauge** is used.

Many different kinds are made, but those generally used agree in the main details. They consist of a piece of wood, the *stock*, which has at least two parallel plane surfaces. A *spindle*, either circular or square in cross-section, passes through a mortise in the stock. At one end of the spindle is a sharp lancet-shaped steel *marker*. Some Swedish marking gauges have two spindles. That side of the stock which is placed against the edge to which the lines drawn are to be parallel, may vary in length, but when lines are drawn parallel to a straight edge (the most usual case), the longer the stock is the better, because this facilitates the accurate management of the tool, and enables even an inexperienced hand to gauge.

(1.) **Marking gauge** with rectangular long stock and cylindrical spindle (Fig. 20). The stock is sawn into at one end as far as the mortise, and to secure the spindle after insertion this end is furnished with a screw, by means of which the spindle is held fast in

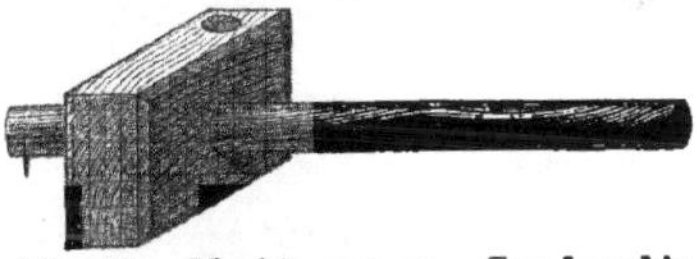

Fig. 20. Marking gauge. Lundmark's patent. 1/4.

* Where the English system of measurement is followed, a *two foot rule* is used, divided into eighths of an inch on one side, and into sixteenths on the other. The use of the metre-measure is, however, strongly recommended. (See footnote, page 13).—Trs.

the manner indicated in Fig. 20. If a thumbscrew and nut are substituted for this screw, the necessary pressure can be more easily and surely produced. (*See* Plate X).

(2.) **Marking gauge** (Fig. 21) with rectangular long stock and rectangular spindle. The spindle is held in place by wedges. This is a simple and inexpensive marking gauge, invented by Herr Alfred Johansson, head-teacher at Nääs. It is recommended as a useful and practical tool for school purposes. (*See* Plate X.)

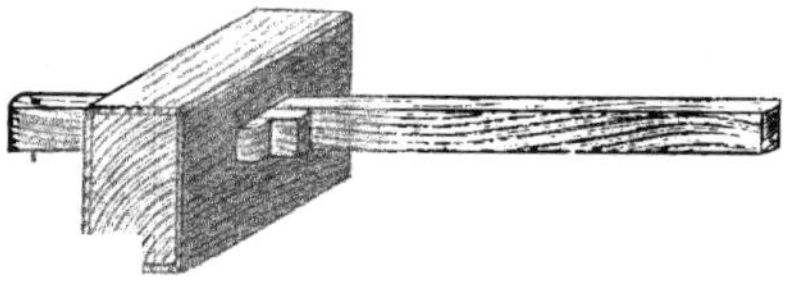

Fig. 21. Marking-gauge. 1/4.

The long stocks of both these marking gauges give them the advantage already indicated over those hitherto in use, *i.e.*, they enable inexperienced workers to gauge without difficulty.

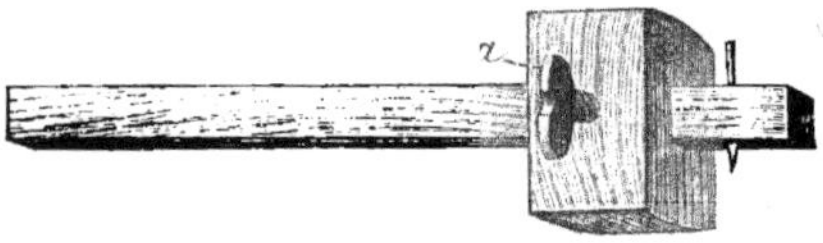

Fig. 22. English Marking-gauge.
a. Thumbscrew.

The English marking gauge (Fig. 22) differs from the Swedish one in having a thumbscrew *a* on one side of the stock, which works against the spindle and holds it in position.

The Marker. The marker must be kept well filed and pointed to secure fine distinct lines, parallel throughout with the edge. The side farthest from the stock should be straight, and as nearly as possible parallel with the side of the stock. The inner side of the marker, on the contrary, should be slightly convex. The marker is thus calculated to cut inwards *away* from the edge, and does not "run off the lines" as a bad marker does, when it meets with a hard layer of autumn wood in cutting in the direction of the grain. With a good marker the gauge should act easily and well without exertion of any kind on the part of the worker.

(3) **The cutting gauge** has a parallelopiped shaped spindle secured by a wedge (Fig. 23). Instead of a pin-shaped marker it is provided with a thin steel cutter, adjusted by means of a pin. Cuttings more or less deep may thus be made on the surface of the work. This tool is chiefly used for gauging across the grain, and in setting out for grooving and dove-tailing.

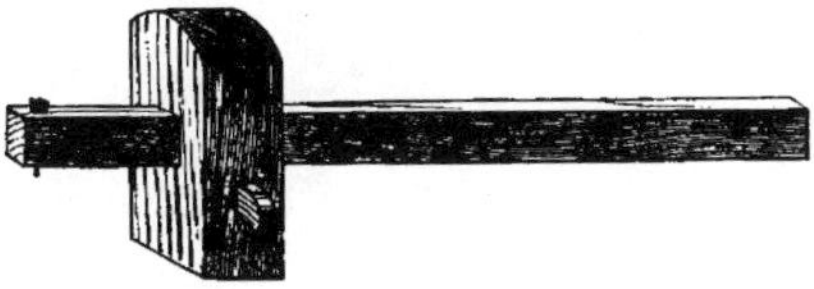

Fig. 23. Cutting Gauge. $\frac{1}{4}$.

In this, as in all marking gauges, it is important that the marker should be inserted in such a way that the inner side, and consequently the point, is slightly inclined outwards from the side of the stock.

IV. Compasses.

1. The compass generally used in slöjd is a simple one made of steel with a hinge. As it is often necessary to maintain the distance between the arms unaltered, this compass is provided with a bow, which is attached to one arm, and which can be secured to the other by a screw. A compass of this kind is called a **bow-compass** (Fig. 24).

Fig. 24. Compasses. $\frac{1}{4}$.

When segments of large circles have to be described, **beam-compasses** are used. In place of the arms of the ordinary compass, these are furnished with trammels, *aa*, united by a cross-piece or beam, *b*, and pointed at one end, where there is a steel pin. One of the trammels is fixed to the cross-piece; the other is movable, and is adjusted by means of a pin.

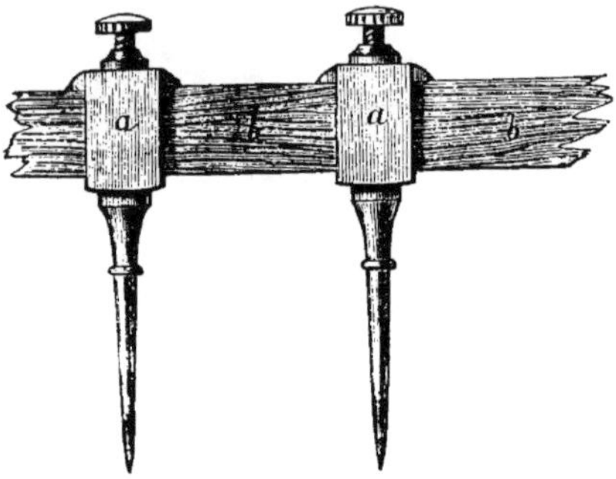

Fig. 25. Trammel Heads, or Beam Compasses.

aa. Trammels, *bb*, Beam.

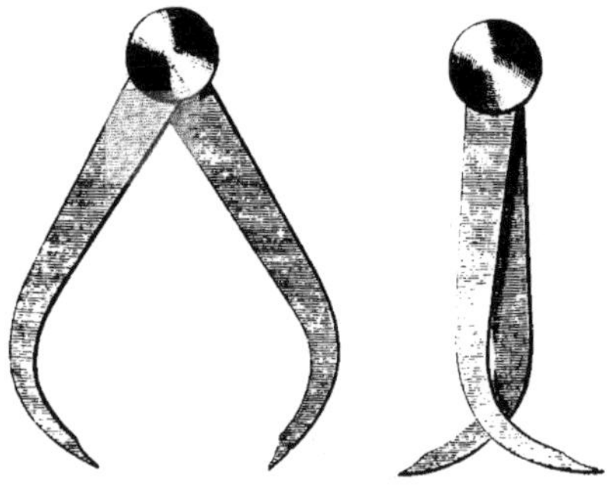

Fig. 26. Caliper Compasses. $\frac{1}{3}$.

2. **The Caliper Compass** is used to measure the thickness of round or oval objects. This compass has very strong curved arms with points which taper obliquely. The ordinary caliper compass may be used to measure the diameter of a hole, by turning the arms round the hinge until the points are turned away from one another (Fig. 26).

V. Squares and Bevels.

Squares are used for testing right angles, bevels for testing angles of various sizes.

Wooden squares.

The Square consists of a short thick piece called the *stock*, with a longer, thinner piece at one end, and at right angles to it, called the *blade*. The stock projects beyond the sides of the blade, and the tool can be easily applied to the straight edge of a piece of wood, that lines may be drawn on the surface at right angles to this straight edge. All the angles of the square, exterior as well as interior, must be perfect right angles. This is not only essential for the operation just described, but also because the square is used for testing solid angles, *e.g.*, the edge of a plank, a corner, etc.

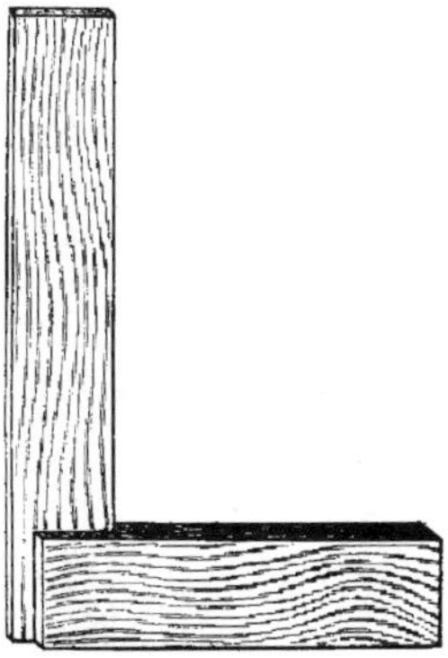

Fig. 27. Wooden Square. ⅛.

Every good collection of tools should include several squares of different sizes, *e.g.*, with blades 6, 8, 12, and 18 inches long.

Steel squares.

The square should be made of hard well-seasoned wood, warranted not to warp. To give greater durability the blade is often made of steel, and the wood of the stock faced with brass on the inner side (Fig. 28). Still stronger and more trustworthy squares are made with steel blades and cast-iron stocks. Squares of this kind are particularly useful as *testing squares*, and one ought to be included in every good collection of tools.

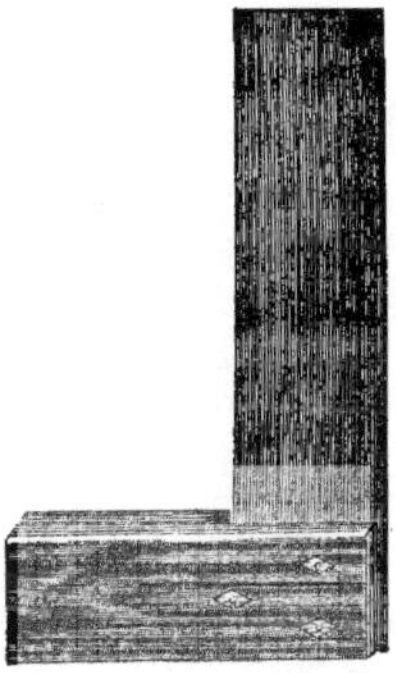

Fig. 28. Square with steel blade. ⅛.

To test a square. The blade is laid on the plane surface of a block of prepared wood, with the stock against a perfectly straight edge. Lines, drawn against each side of the blade, are then made on the wood. The square is next reversed, the stock is placed as before, and the edges of the blade are placed close to the lines previously made. Lines are then drawn once more along the edges of the blade. If these lines coincide, or are perfectly parallel with those made first, the square is *correct.*

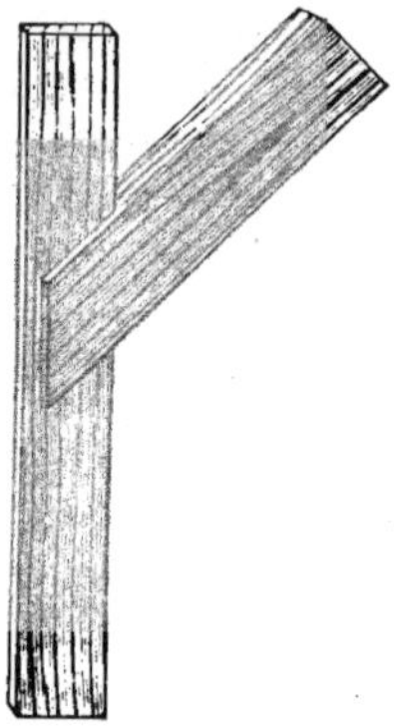

Fig. 29. Set-bevel or mitre-bevel. $\frac{1}{5}$.

2. The **set-bevel** (Fig. 29) consists, like the preceding, of a *stock* and a *blade,* but the latter, which generally extends beyond the end of the stock, is attached in such a way that it forms on one side an angle of 45°, and on the other an angle of 135°, or the complementary angle of a straight angle. It is used when a rectangular corner is made by joining together pieces cut at an angle of 45°. Such pieces are said to be *mitred.*

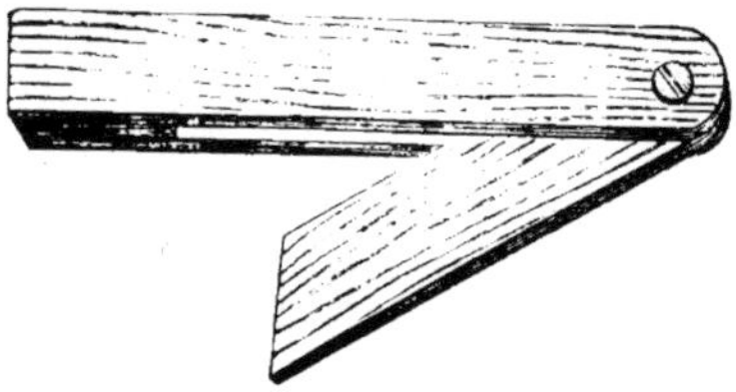

Fig. 30. The Wooden Bevel. $\frac{1}{6}$.

In the **wooden bevel** the blade rotates on a screw in the stock. To secure the blade in any given position the screw is furnished with a nut, by means of which it may be screwed fast. (Fig. 30.)

To test a plane surface.

VI. Winding laths or **straight edges.** To test the accuracy of plane surfaces, a long, perfectly straight ruler or *straight edge* is used. When this is placed on the surface in various directions, there must be complete contact between it and the surface. A still more delicate method of proof is furnished by the double straight edge, or two straight edges exactly the same (Fig. 31). In applying the test the straight edges are placed one at each end of the piece of wood, and parallel to one another. On careful

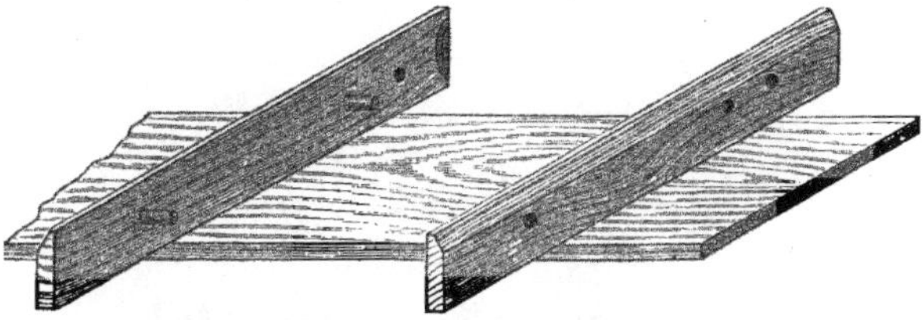

Fig. 31. Winding laths or straight edges. ⅛.

inspection, if the surface is level the upper edges of the rulers will be found to be in the same plane. The straight edges, when not in use, are held together by a couple of pegs.

The edge of the trying-plane is often used instead of the straight edge, and two trying-planes instead of the double straight edge. See further under "face planing," p. 132.

D. Tools used for cutting up wood and making the articles.

I. Saws.

The saw is an indispensable tool, and in the case of most articles it is the first used. The *blade* is made of thin steel of various breadths, on one edge of which a series of sharp points form the *teeth.* The steel must be soft enough to be acted on by the file, and to admit of the teeth being slightly turned aside without breaking off.

The saw acts by tearing or cutting the fibres of the wood as the teeth of the blade pass over them. The teeth are, therefore, the characteristic part of the saw, and its efficiency depends on their form, size, and quality.

The **shape and size of the teeth** vary considerably in different kinds of saws. The form generally used in wood slöjd is shown in Fig. 32. The form of the teeth is that of

Fig. 32. Teeth of a bow saw for ripping. $\frac{1}{1}$.

a scalene triangle, the base of which is formed by the blade. The shortest side froms an angle of 80°-90° with the base. In the frame saw (Fig. 37, *B*), the angle is 90°; in the bow-saw, the dove-tail saw, etc., it is 80°-85°.

Form and position of the teeth of a saw.

The teeth of any given saw must always be alike in size and shape, and must always be *set at the same angle.* The shorter side of the teeth, being nearly at right angles with the blade, is the cutting side, and in working the saw this is the side which should enter the wood. When the saw is drawn back, the more sloping side of the teeth has very little effect upon the fibres, and the saw "goes empty."

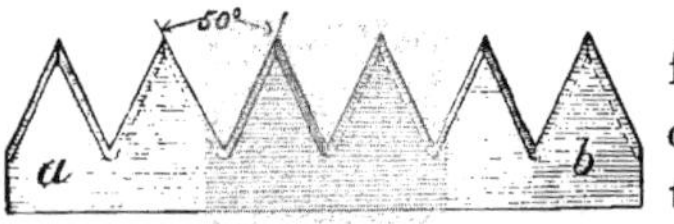

Fig. 33. Teeth of a bow-saw for cross-cutting, or wood-saw. $\frac{1}{1}$.

The teeth of the bow-saw for cross cutting form an isosceles triangle of 50° between the teeth. A saw of this description cuts equally well backwards or forwards

Length of the teeth.

The space *between* the teeth must be great enough to leave room for the *sawdust* until the saw has carried the latter beyond the wood. Now, as the sawdust occupies more space than the wood from which it is produced, the teeth of the saw must be considerably longer than the depth of the cut made *each time* the saw passes through the wood, and the point only of the teeth must be allowed to cut the wood, to prevent hindrance to their action by an accumulation of sawdust. If the sawdust prevents the free passage of the saw, or if it clings about the teeth, it is either because the teeth are too small, or because too much pressure is laid on the saw.

Why the saw must be set.

It is almost impossible to avoid considerable friction between the blade and the sides of the cut, and this friction is increased by the sawdust which accumulates at the sides of the blade. It is therefore necessary to give the blade a certain amount of "play;" in other words, *the breadth of the cut must be greater than the thickness of the blade.* This is effected by bending the teeth alternately a little to the one side and to the other, or, as it is termed, by *setting* the saw.

Setting is performed by means of the **Saw-set**, a steel blade $\frac{1}{16}$ inch thick, the edges of which are indented by notches of various breadths.

Fig. 34. Saw-set. $\frac{1}{8}$.

Some English Saw-sets are furnished with an adjustable slide rest. In setting a saw the blade is fastened into **Saw Sharpening Clamps** (Fig. 35) and these are screwed to the bench. One tooth after another is grasped by the notch of the Saw-set best adapted to the thickness of the tooth, and the blade of the Saw-set being held in such a way as to conceal the point of the tooth, the latter is then *turned sharply aside.* It must not, as is sometimes done, be twisted at the same time in the direction of the length of the blade, as this may cause it to break off. Great accuracy is required in the operation, and the setting should never be so extreme that the width of the cut is more than double the thickness of the teeth. If this width is exceeded the saw will not act easily.

To what extent a saw should be set.

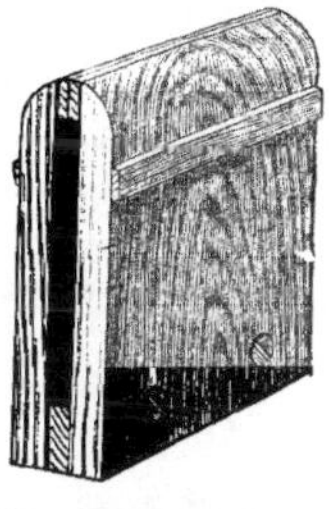

Fig. 35. Saw sharpening clamps. One half loosely fastened to the other by means of wood screws. $\frac{1}{8}$.

Unevenly set saws.

Considerable practice and skill are required to set a saw. The points of the teeth should form a line exactly parallel to the length of the blade, but it often happens that some teeth project beyond this line and others fall within it. This fault may be remedied to some extent by drawing the blade between a couple of gouges, fixed points downwards in a piece of wood, with the convex sides facing one another.

The blade of the saw is placed between them, teeth upwards, and the points turned from the operator, or in the direction from *d* to *c* (Fig. 32), the handles are grasped with one hand to bring the gouges close together, and the blade of the saw is drawn forwards between them.

Setting tongs. In consequence of the difficulty of setting a saw evenly and at a good angle, many different kinds of saw-sets and setting-tongs have been devised. The latter are intended to be adjustable for any desired inclination of the teeth. Some of these tools, however, are not practically useful, and those which are fully adapted for use are generally too expensive for ordinary purposes.

As indicated above, *setting* must not go beyond a fixed limit. Provided that the saw has free passage through the wood, the finer the cut it makes the better; and much less inclination of the teeth is necessary, in the case of dry timber, than in unseasoned or loose-fibred wood.

Saw blades with thin backs. Less setting is also necessary in the case of saw-blades which increase in thickness towards the teeth. These are made in the best manufactories, and are always preferable to blades of equal thickness throughout. So-called *compass saws* often have blades of this kind, and require no setting.

Sharpening the saw. Quite as important as the setting of the saw is its *sharpening*, and it is often necessary to perform both operations at the same time.

To **sharpen** a saw, it is secured in the saw-sharpening clamps; and the ordinary kinds of saw used in wood slöjd are sharpened by means of a **triangular file** (Fig. 36).

Fig. 36. Triangular or Three-square File. ½.*

Care must be taken that the two sides of the file which

* The file represented in the illustration is a single-cut file; but a double-cut file should be used.—Trs.

are to be used form the angle necessary to produce the inclination in the edges of the teeth indicated above. This being secured, *the file is drawn across the blade at right angles to it.* Every indentation must be filed equally deep, or, in other words, the point of each tooth must stand equally high. The row of teeth is next tested with the straight edge, and if any of the teeth stand higher than the others, they must be *topped* or *filed down* with a fine broad file, and then sharpened once more.

Sharpening is begun at the end of the blade, towards which the points of the teeth are turned, or from *c* to *d* (Fig. 32). The degree thus produced on the points is always in the direction to which the teeth are turned, *not away from it.* In the latter case, the saw would be rather blunt. Each tooth must be carefully filed, that its edges may be quite *sharp*, and the cutting side quite *straight.*

Should the saw, after sharpening, be insufficiently set, it must be set again, after which the file must be once more passed over the teeth to remove any irregularities. Generally speaking, setting precedes sharpening.

Obliquely sharpened teeth.

Sharpening is sometimes performed by passing the file obliquely over the edge of the blade, instead of at right angles to it. The edges of each tooth are thus sharpened obliquely from within outwards (see Fig. 33). The file is first passed obliquely through *every alternative* tooth-space. The saw is then reversed, so that its ends change places, and the remaining spaces are operated on in the same way. This gives a knife-like edge to both sides of the teeth, and makes the saw cut particularly swiftly and well. The common *wood-saws*, some *tenon-saws*, and *hand-saws*, are sharpened in this way.

It need hardly be added that setting and sharpening are not only necessary in the case of new saws, but also as often as the teeth become worn or blunt.

The saws now to be described may be classed in two groups, *i.e., saws with frames, and saws without frames.*

The former have the ends of the blade fastened into a frame, the tension of which may be regulated to produce the necessary amount of resistance. In the latter kind of saw this power of resistance is given by means of the greater breadth and thickness of the blade, or by setting the back of the blade in a binding of metal. This binding is called the saw-back.

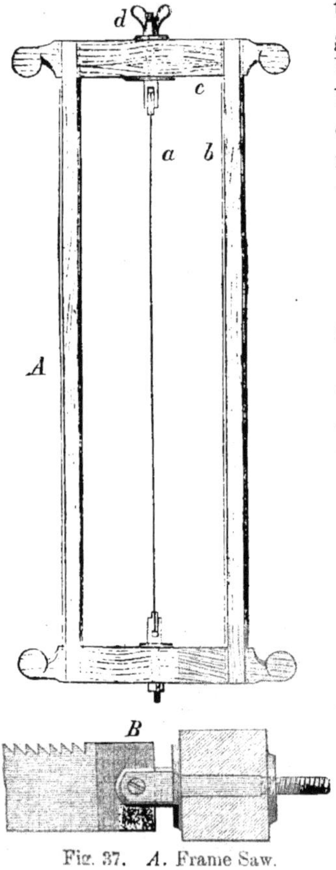

Fig. 37. *A*. Frame Saw.
a blade. *b* side-rail. *c* top-rail. *d* winged nut and saw-blade attachment. $\frac{1}{18}$.
B. Saw blade end with attachment. $\frac{1}{6}$.

1. Saws with Frames.

1. The Frame Saw (Fig. 37) is the largest saw used in Slöjd. It is used for sawing up planks and other pieces of wood lengthwise into thinner pieces. It is worked by two people, and in a horizontal direction. The blade has from 3 to 4 teeth per inch, and it is fastened into an oblong wooden frame, midway between the side-rails.

The ends of the blade are enclosed in and strengthened by pieces of white-iron, and are fastened by the attached pieces running through each *top-rail*. Tension is produced by turning the winged nut. The cutting side of the teeth is at an angle of 90°.

2. **Bow Saws** (Fig. 38) are of different sizes. They are much used in wood-slöjd, not only in the earlier, but in the later stages of work. Bow saws have all the same kind of frame, consisting of a bar called the *stretcher*, longer than the blade and parallel to it, at each end of which there is either a square mortise or a fork-like notch for the reception of the cross-pieces or *side-arms*. The latter, though carefully fitted in, yet have a certain amount of play at the ends of the stretcher, in order that they may be drawn closer to each other on either side of the stretcher when the saw is tightened. At one end of each side-arm there is a round hole, through which passes a well-fitting *peg* with a handle. This peg is sawn through the middle lengthwise to form a slot for the saw blade, which often extends a certain length into the handle. The blade of the saw is narrower at the ends where it enters the handle. In it are one or two holes, through which the fastening pin runs.

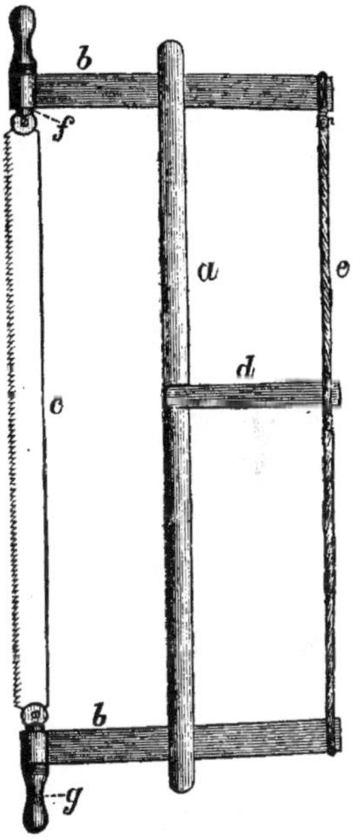

Fig. 38. Broad-webbed Bow-saw.
a stretcher. *bb* side-arms. *c* blade. *d* tightener. *e* string. *f* end of blade with attachment. *g* handle. 1/12.

Blades fastened in this way often twist when tightened, and consequently cut badly. This happens especially when the axis of the handle is not exactly in line with the blade. This defect may be remedied by substituting for prolongations of the blade itself, the white-iron attachments (Fig. 39), and securing them in the usual way. The ends of the blade are fastened between the plates of the attachment merely by a screw or nail, in order that the blade may be freely adjusted.

The side-arms are connected at the other end by several strands of strong string, which are twisted together by a *tightener*, in order to give the required tension to the blade. When the string is put on, the frame is fastened between the bench pegs.

The stretcher is made of fir or pine; the side-arms of harder wood, *e.g.*, beech or oak. The different parts of the frame are made as light as is compatible with strength, that the saw may not be too heavy to manage with one hand.

Manner of holding the saw.

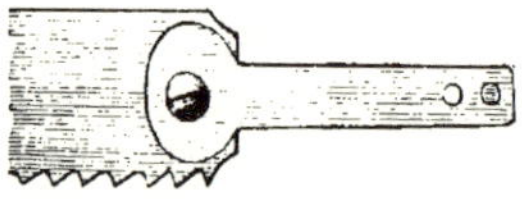

Fig. 39. Saw-blade end, with attachment. ¼.

In working, the saw should be firmly grasped by the side-arm just above the handle. In the case of the lighter description of saws, the handle, as well as the lower part of the side-arm, should be held in the hand, and the index finger should steady the blade.

Generally speaking, the blade is fixed obliquely to the plane of the frame; partly that the worker may saw deeply without hindrance from the frame, and partly that he may be able to see the line which the saw is to follow.

Tightening the blade.

In tightening the blade—which is best done by turning both handles simultaneously—care must be taken that it is perfectly straight. Otherwise a straight cut can hardly be obtained.

If the saw is out of use for any length of time, the tightener should *always be slackened.* When this is not done the side-arms may become twisted.

Bow-saws have different names, depending on the nature of the blade. The "hook," *i.e.*, angle of the teeth is shown in Fig. 32.

A. **The Broad-webbed Bow-saw** is shown in Fig. 38. Its blade is 1 to $1\frac{1}{2}$ inches broad. It is used in numerous cases, *e.g.*, in sawing off long slips of wood, where a straight cut is all that is required. It has 4 to 5 teeth per inch.

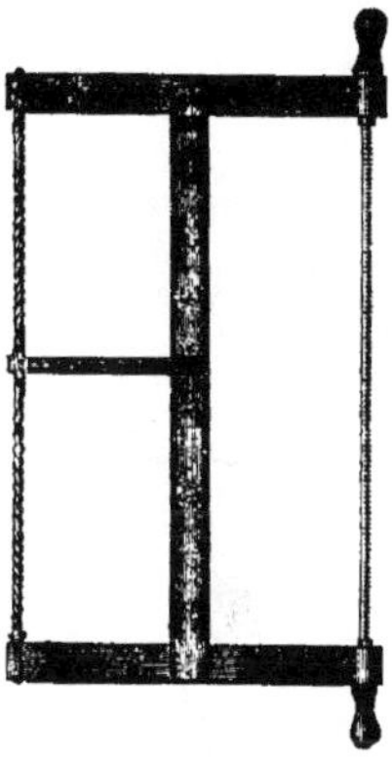

Fig. 40. Turn-saw. $\frac{1}{12}$.

B. **The Turn-saw** (Fig. 40). The frame resembles the preceding, but the blade is very narrow—about $\frac{1}{4}$ inch, or very little more—because it is used to produce curvilinear cuts. The toothing is very fine—7 teeth per inch—and the setting is sometimes less than in the bow-saw, that the cut may be accurate, and not unnecessarily broad.

Turn-saws, the blades of which are over half an inch in breadth, are also used. These are called *broad-webbed turn saws.*

2. Saws without Frames.

1. **The Hand-saw** (Fig. 41) has a very broad blade, which is narrower at one end, and is provided at the broader end with a convenient handle. The large blade gives it sufficient strength, and this is often increased by the thickness of the blade, which may exceed that of the frame-saw. The teeth are set to cut when the worker pushes the saw away from him, but not when the saw is drawn back.

This saw, distinguished for its simplicity and convenience in working, is in general use in England and North America, but is not much used in Sweden.

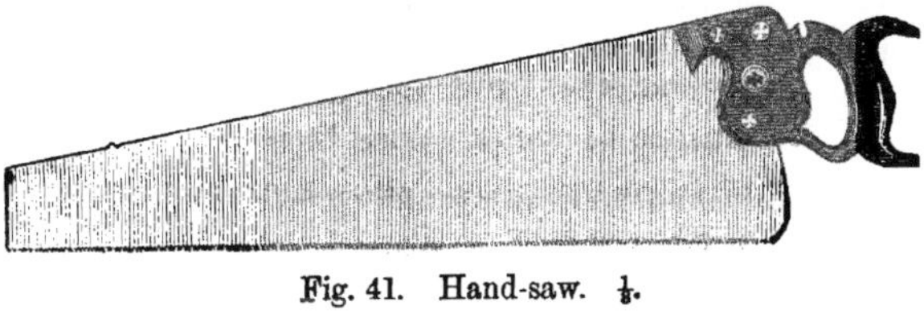

Fig. 41. Hand-saw. $\frac{1}{8}$.

2. The **Dovetail saw** (Fig. 42) has a very broad blade of equal breadth throughout, with a handle. To give sufficient strength to the blade, its upper edge is enclosed in an iron back. This thick back limits the depth of the cut; consequently this saw is only used for shallow incisions, *e.g.*, in sawing out tenons, dovetails, etc. This saw has 10 to 12 teeth per inch. The shape of the teeth is shown in Fig. 32, but they are often sharpened with advantage in the manner shown in Fig. 33.

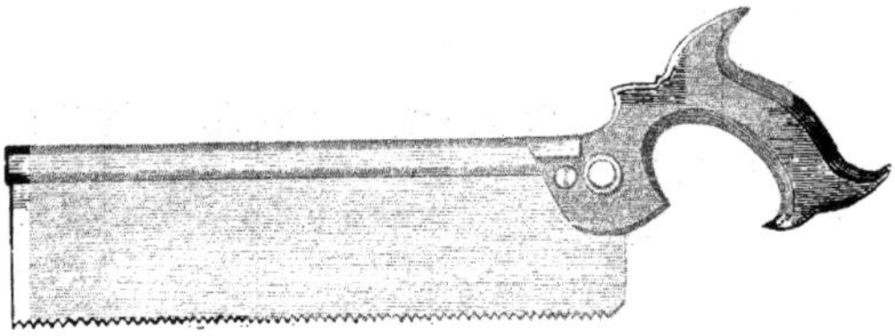

Fig. 42. Dovetail-saw. $\frac{1}{8}$.

[3. The **Tenon-saw** is practically the same as the dovetail-saw, but it is rather larger, and it has what is called a Box-handle, somewhat like that of the hand-saw.—TRS.]

4. The **Compass-saw** (Fig. 43). The blade is very narrow, and terminates in a point. This saw is used when an excision has to be made in the centre of a piece of work, and cannot be begun from the edge. For this purpose a hole must be bored, into which the point of the saw can be inserted. To give the blade sufficient strength it is made tolerably thick, but it becomes thinner towards the back. Compass-saws are of various sizes, and the teeth are set in different ways. The number of teeth varies from 5 to 12 per inch, but their form is in most cases that shown in Fig. 32.

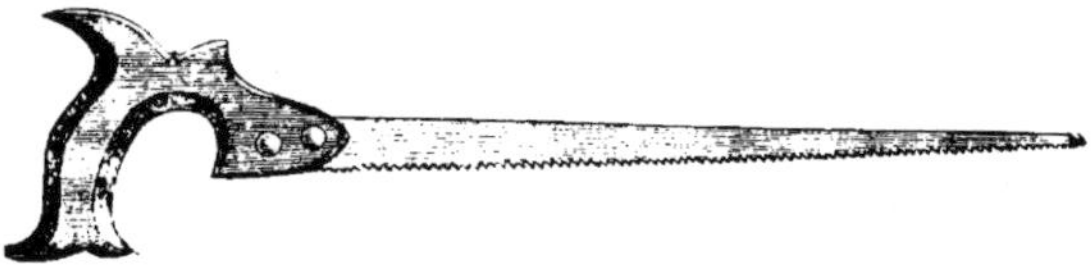

Fig. 43. Compass-saw. $\frac{1}{6}$.

4. **The Groove-saw** * (Fig. 44) has a tolerably thick blade

* Unknown in England, but recommended as useful.—TRS.

of equal breadth throughout, the upper edge of which is entirely enclosed by a handle, which is worked by both hands. The teeth are inclined towards the worker, and consequently act when he draws the saw towards him.

It sometimes happens, especially in clamping and grooving, that an incision must be made in a broad flat piece of wood, and in many cases it must not be carried to the edge. With the exception of the tenon-saw, the saws hitherto described cannot be used for this purpose. The groove-saw is perfectly adapted for it, whereas the tenon-saw is not quite so convenient, because the setting of its teeth is not suitable, and it has only one handle.

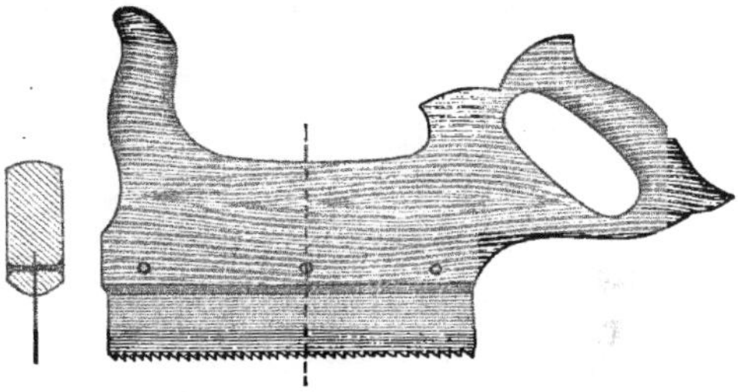

Fig. 44. Groove-saw. $\frac{1}{4}$.

II. The Axe.

After the saw the *axe* is one of the most useful tools in the earlier stages of any piece of work. Axes are of various kinds, manufactured for different purposes. An axe of American construction, very suitable for slöjd work, is shown in Fig. 45. The edge and faces are slightly curved, and ground on both sides. The axe should not weigh more than about 2 lbs., that it may, without trouble, be wielded by one hand. The *handle*, of hard and tough wood, such as oak or ash, should be curved so as to fall well into the hand, and the axe

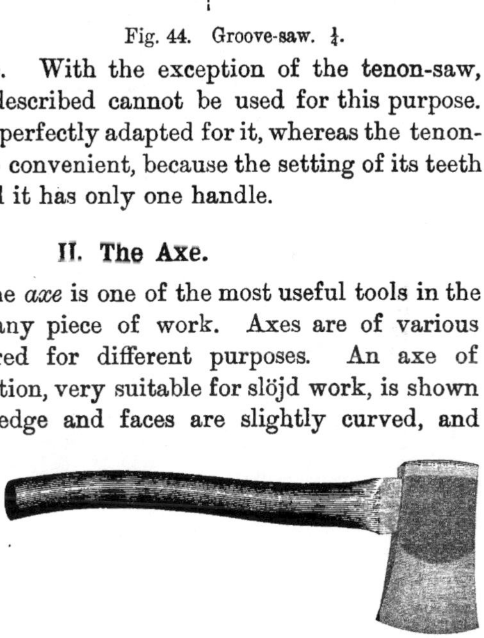

Fig. 45. Axe. Ohio pattern. $\frac{1}{7}$.

shaft must be firmly secured by wedges into the *eye* of the axe-head.

In working with the axe the wood is supported on a *block*, formed of an evenly sawn-off piece of the trunk of a tree. The best tree for this purpose is the poplar.

The surface of the block must always be kept free from sand, which would destroy the edge of the axe.

It is of the utmost importance for beginners to hold the piece of wood in such a way that the hands may receive no injury.

Grinding the axe.

In **grinding** (see under this head, pp. 115-118) the axe and all other edge-tools, the tool must be held steadily against the grindstone, in order that the bevelled edge may be quite regular and of the same breadth, not waving. The two bevelled edges should form an angle of about 20°.

III. The Knife.

A suitable slöjd knife.

The *knife* is the *slöjder's* indispensable and most important tool, and it is the *first* to be placed in the hands of a beginner. It is therefore important to select for slöjd suitable knives of the best quality. The blade of the slöjd knife should be made of good steel, about 4 inches long, and not more than $\frac{3}{4}$ inch broad. The edge should be straight, and the two faces which form it should extend over the entire breadth of the blade. The back of the knife should not be more than $\frac{3}{16}$ inch thick. The blade ought not to taper to a dagger-like point, but should terminate as is shown in Fig. 46. The best angle for the edge is 15°. The other end of the blade terminates in a *tang* which slots into the handle.

Fig. 46. Slöjd-knife—Nääs pattern. $\frac{1}{3}$.

A commoner, though by no means so suitable form of knife is shown in Fig. 47. Directions for using the knife are given in Chap. V.

Fig. 47. Slöjd knife. $\frac{1}{3}$.

IV.—The Draw-Knife.

This consists of a steel blade with an edge formed by grinding on one side only. This blade is furnished at both ends with handles, at right angles to it, and in the same plane. The tool is worked with both hands, so that the whole strength of the slöjder can be thrown into its use. The draw-knife is chiefly used in modelling and smoothing objects with curved outlines. It is also used in making hoops for barrels, &c. Directions for its use are given in Chap. V.

Use of the draw-knife

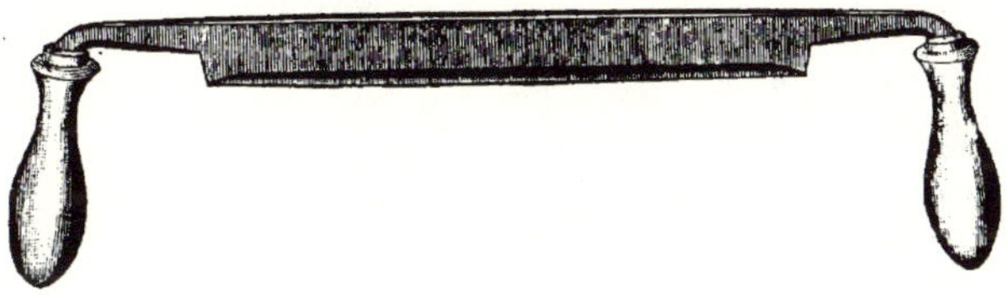

Fig. 48. Draw-knife. $\frac{1}{4}$.

V.—Chisels, Gouges, Carving Tools, &c.

These terms include a whole group of tools which are used in wood-slöjd for the removal of small pieces of wood, in cases where the knife, the saw, or the plane could not advantageously be used.

Parts of a chisel, &c.

They consist of a flat or concave blade made of steel, the cutting end of which is cut straight across and sharpened to an edge, and the other wrought into a four-sided *tang*, which is set into a wooden *handle*. The tool in working is driven into the wood either by the pressure of the hand, or by blows from a mallet. In order that the handle may not slip, or twist round when grasped, it is generally made with four sides, greater in breadth than in thickness, and with the broader sides rounded.* To keep the handle from splitting under violent pressure, the base of the tang is furnished with a *shoulder*, on which the handle rests.

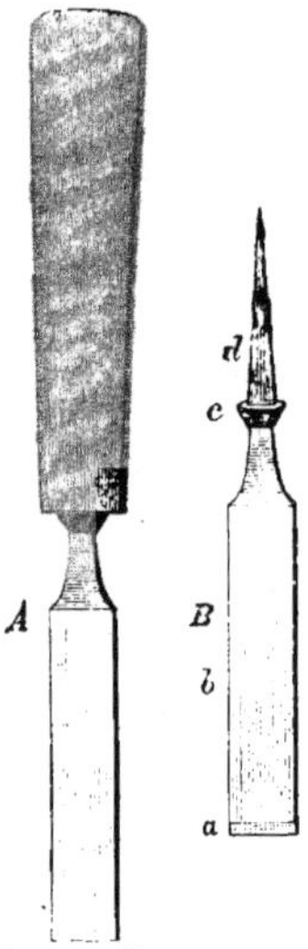

Fig. 49. Firmer Chisel ¼.
A. Blade and handle.
B. Blade showing a face and edge.
C. Blade. *c*. shoulder. *d*. tang.

These tools vary greatly in size both as regards length and breadth. The latter dimension determines the dimensions of the edge. The broadest tools are generally also the longest.

In order to be able to execute all the different kinds of exercises which occur, it is necessary to have a complete set of each description of tools. There are usually 12 in a set, all of different breadths.

Tools of this kind are classified according to the different shapes of the blade and edge, and the different methods of sharpening as follows :—

1. Chisels.

These tools have a *straight* edge ground on *one* side.

1. The **Firmer Chisel** (Fig. 49). The breadth of the blade, which varies from 1½ inches to ¼ inch, is generally much greater than its thickness. The face of the edge in all such tools forms with the front side an angle of 20° to 25°.

The firmer chisel is used in paring plane or convex surfaces; in mortising, when it often does duty instead of the mortise

* English handles are generally *turned* in boxwood or beech.—TRS.

chisel; in curved work; in facing off; and, generally speaking, in all cases where no other tool can be made use of with advantage.

2. The **Mortise-chisel** (Fig. 50). The thickness of the blade generally exceeds its breadth, which varies from $\frac{1}{8}$ inch to 1 inch. The front face of the blade is always a little broader than the back.

The mortise-chisel is used for mortising; and, whenever possible, a blade of the same breadth as the mortise to be made should be selected. The great thickness of the tool enables its sides to act with steady force upon the sides of the mortise, and makes accurate execution of the operation much easier. It is driven into the wood by blows from a mallet. The angle of the edge is the same as in the firmer chisel.

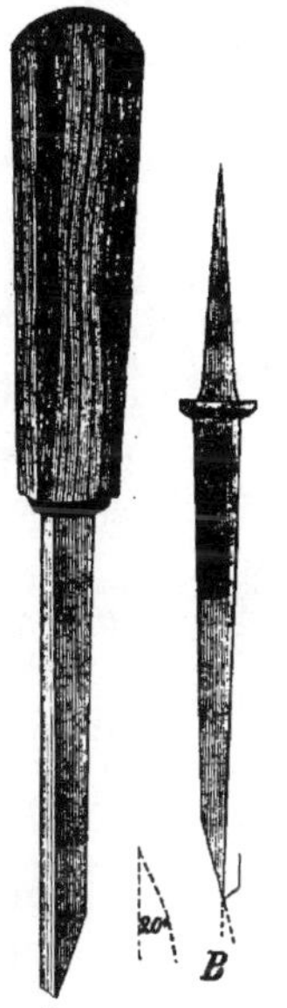

A

Fig. 50. Mortise-chisel. $\frac{1}{4}$.
B. shows breadth and angle of the edge.

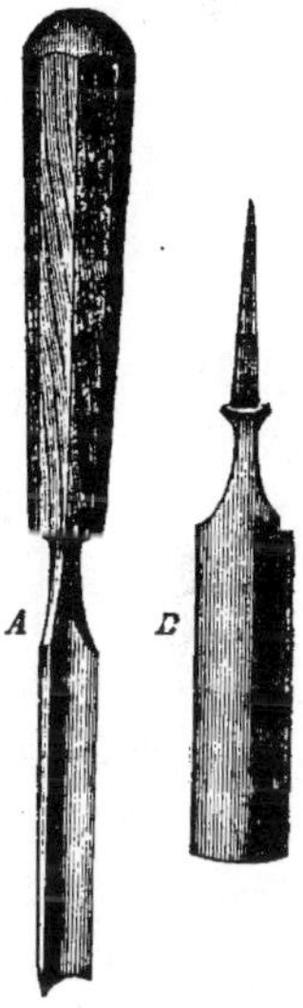

Fig. 51. Gouge $\frac{1}{4}$.
A. with edge on the inner side.
B. with edge on the outer side.

2. Gouges.

These tools have a *curved* edge.

The blade of the gouge is concave. The face of the edge may either be (*a*) ground from *within outwards,* in which case the edge will lie upon the inner or concave side, or (*b*) *in the reverse way,* when the edge will lie upon the outer or convex side.

Gouges ground in the first mentioned manner are used in the formation of grooves or bowl-shaped depressions. Those ground in the other way are used chiefly in perpendicular paring to produce concave and cylindrical surfaces.

The breadth of the gouge varies from $\frac{1}{4}$ inch to $1\frac{1}{2}$ inches, and the curve of the edge may include from one-tenth to one-half of a circle, or 36° to 180°. All the gouges in one set should have the same curve in the edge. The gouge is driven into the wood by the hand, or in the case of gouges of large size, by the mallet.

3. The Spoon Gouge and the Spoon Iron.

Ordinary gouges are often used in forming the bowls of spoons and similar articles, but the tools specially adapted, and best for the purpose, are the *spoon gouge* and *spoon iron.* The larger illustration (Fig. 52) shows the spoon gouge. In construction, and in the way it is used, it somewhat resembles *A* (Fig. 51); but it differs from it in having the blade curved lengthwise, to facilitate the work of hollowing out.

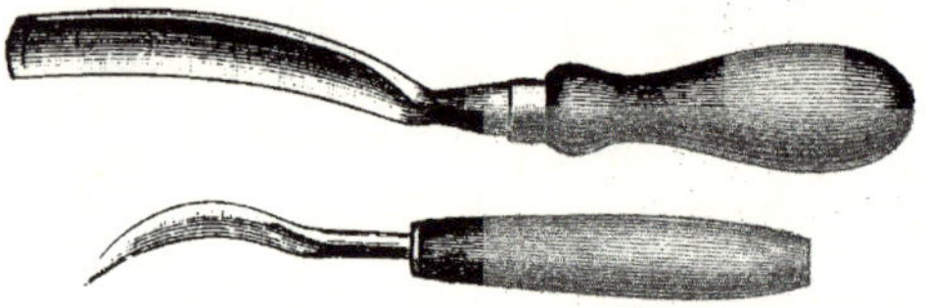

Fig. 52. Spoon Gouge and Spoon Iron. $\frac{1}{4}$.

The spoon iron is different in form. It is shown in the smaller illustration (Fig. 52), and resembles a knife with a lancet-shaped blade, with two edges, curved like a bow, and tapering to a point at the end. It is worked with both hands, and cuts to either side.

4. Carving Tools.

A number of tools, more or less like the preceding, are used in wood-carving. Some of these carving tools are flat, with rectangular edges; others are oblique to the direction of their length, with a bevelled edge on both sides; others are concave, with a circular edge, or have two edges meeting in a point. They are straight in some cases; in others, curved.

As only a few of these tools are used in slöjd carpentry, to any extent worth mentioning, no description of them is given; but those in most common use, with their names, are shown in Fig. 53. The full size of the edge is given in the outline beside the representation of each tool.

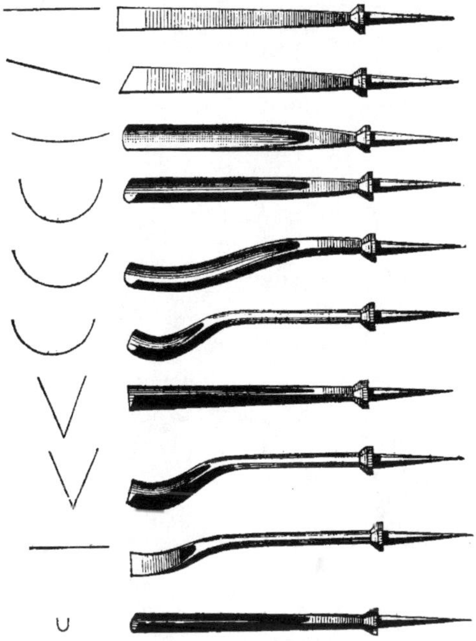

Fig. 53. Carving Tools. ¼.

VI. Planes.

The edge tools hitherto described consist of a single steel blade, with a cutting edge of various descriptions, and a handle for one or both hands. The inclination of the edge to the surface of the wood may thus be altered at will, as the circumstances of the case require. Narrow surfaces, or surfaces of generally circumscribed area, may thus be levelled and smoothed to a certain extent (though not perfectly) by the knife, the axe, the chisel, etc.; but when long and broad surfaces have to be made absolutely smooth, we require an edge-tool which, by attacking in the first place all the elevations, and by always cutting equally deep on a plane surface (*i.e.*, by always removing shavings of the same thickness), finally reduces the surface to one uniform level.

The **plane** is the tool which fulfils these requirements. In the plane, the steel blade called the *plane-iron* is wedged tightly into a parallelopiped-shaped wooden block, called the *plane stock*, which is formed in various ways for various purposes. The edge of the blade extends slightly beyond the under side of the block.

The plane is used not only in the dressing of *plane* surfaces, but also in the preparation of all surfaces on which straight lines can be drawn in at least *one* direction; *e.g.*, in smoothing the surface of cylindrical and conical objects, etc. Consequently, many different kinds of planes are required.

The plane-stock. **All planes,** however, consist of two principal parts: the *sole* or *stock*, and the *iron*. The stock is formed of hard, tough, straight-fibred wood in the form of a parallelopiped, the under side of which, the *sole*, glides over the work when the tool is used. The best wood is elm, beech, pear, or boxwood, which has been well seasoned to prevent warping. The plane is worked with both hands. The front part of Swedish planes is often provided with a rest for the hand, called the *horn*. The larger kind of planes have a handle behind the iron.

The plane-iron. The plane-iron is placed obliquely in a hole in the stock, called the *socket* (Figs. 54 and 56), with its edge extending a very little beyond the sole, and it is secured by a wooden *wedge*. It is made of iron, with a steel front. In shape it resembles a wedge, the thicker end of which is sharpened. The wedge-like shape gives the required thickness and strength to the sharpened end, leaves more room towards the upper end, and also helps to keep the plane-iron firmly in its place when the edge comes against hard knots in the wood and the pressure tends to force the iron upwards

To form the edge, the plane-iron is ground on the posterior or bevelled edge. This forms an angle of from 20° to 25° with the front face of the plane-iron. The former angle is suitable for loose fibred timber; the latter for hard or knotty wood. The edge must not be too thin, for if so, the iron will *fly, i.e.*, become jagged. The iron is generally placed in the socket at an angle of 45° to the plane of the sole, with the bevelled edge downwards. *Angle of the edge of the plane-iron and its position in the stock.*

It occasionally happens, *e.g.*, in small American planes with iron stocks, that the bevelled edge of the plane-iron is turned upwards at an angle of 25° to the plane of the sole. It may also be mentioned, in passing, that in planes manufactured for special purposes, *e.g.*, planing particularly hard kinds of wood, the irons are placed at an angle of 50°, 55°, 60°, or even 90°.

As indicated above, the plane acts by removing thicker or thinner shavings, according as the plane-iron extends more or less beyond the sole. In working with the knife it is always possible to alter the position of the edge in order to prevent its cutting in the same direction as the fibres run, which would tear them, and render the surface uneven. But it is not always possible to guide the stationary plane-iron in this way. Hence in cross-grained wood, or in timber where the fibres lie parallel with the surface, the plane has a tendency to split or tear them, and the resistance offered by the torn fibres is often so great that the plane cannot be driven forward. The fibres also, by their elasticity, tend to drag the iron downwards. To prevent the fibres tearing in front of the iron, provision must be made (1) *for breaking them off at once,* and (2) for bringing at the same time *pressure to bear on them* from above, just over the edge of the iron, by means of which their elasticity may be diminished or wholly neutralised. The *first* object is attained by placing a *cover* above the iron, the effect of which is to break off the fibres as quickly as they are detached; the *second,* by reducing the *set* or opening in front of the iron as much as is compatible with the free passage of the shavings through it. *The cover.*

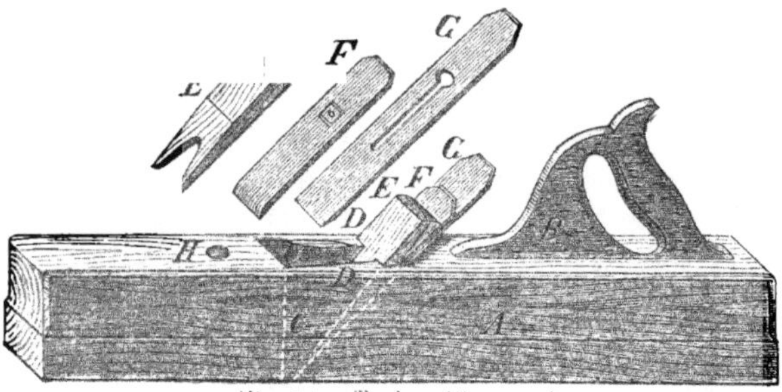

Fig. 54. Trying Plane. $\frac{1}{6}$.
A stock, *B* handle, *C* socket, *D D* cheeks, *E* wedge, *F* cover, *G* iron, *H* boss.

To put on the cover.

A rectangular opening in the iron, enlarged and rounded at one end, admits the screw of the cover, and permits of its adjustment. The lower end of the cover is curved, with the concave side inwards, and it terminates in a sharp edge. When the screw is tightened this sharp edge must lie *close against* the surface of the iron (see Fig. 55). If the slightest space is left the shavings will force their way through. The other side of the cover must be carefully rounded to permit the shavings to glide freely over it. The edge of the cover should be very near the edge of the iron. In finishing up a surface, and plain jointing, the distance should be about $\frac{1}{32}$ inch, and about double that distance in cases where coarser shavings may be removed. The distance between the socket and the edge of the plane in front should be about $\frac{1}{16}$ inch for fine planing, and not more than $\frac{3}{16}$ inch for coarser work.

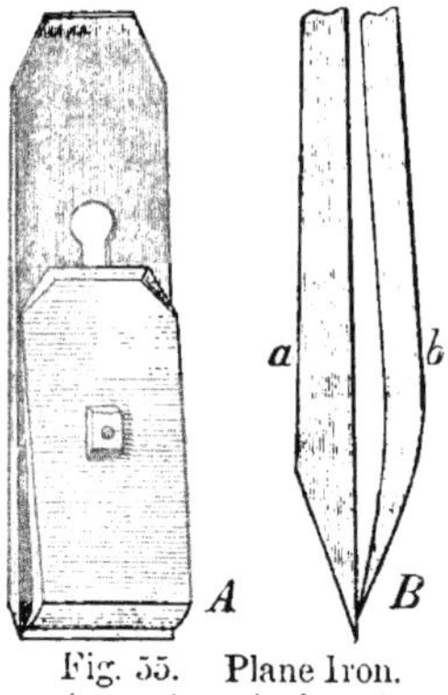

Fig. 55. Plane Iron.
A seen from the front $\frac{1}{4}$.
B seen from the side $\frac{1}{4}$,
a iron, *b* cover.

The wedge and the socket.

In planes like the smoothing-plane and the trying-plane, where the iron is narrower than the sole, and is inserted in the socket from above, the front side of the socket should be at right angles with the plane of the sole, and of the same breadth as the iron. The inclination of the side of the socket on which the iron rests has been already indicated; the other two sides, *i.e.*, the *cheeks*, are thicker towards the iron, in order to give support and steadiness to the wedge, and the sides of the wedge are inclined towards one another at an angle of about 8°. If this angle is much greater the wedge fits loosely; if it is less it may fit so tightly that it cannot without difficulty be loosened. The wedge, which is forked at the lower end, must fit accurately into the space in the socket left by the iron, otherwise shavings may gather round its points (see Fig. 56). These points require frequently to be trimmed, because from repeated sharpening the wedge-shaped plane-iron gradually sinks deeper in the socket, causing the wedge to do the same.

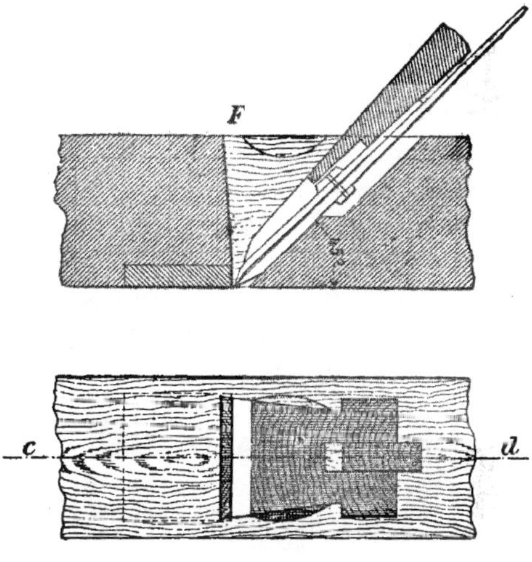

Fig. 56. Portion of Plane. Socket. ¼.
F section through *c d* showing plane-iron, wedge and piece of wood inserted.

Planing the sole and inserting a piece of wood.

Should the sole of the plane become warped, or uneven through wear, it must be carefully planed. It follows from the construction of the socket that the opening in front of the iron, after repeated planing, becomes too large. It is usual to remedy this by inserting in front of the iron a piece of very hard wood, *e.g.*, ebony, beech, or boxwood (see Fig. 56). Brass is also used for this purpose. New planes are also often furnished with such pieces, in order that the portion in front of the plane-iron's edge may longer resist the wearing effect of the shavings.

Putting in the Plane-Iron or Setting the Plane.

The cover is screwed tightly on the iron, with its sharp edge at the proper distance from the edge of the iron, which is then laid in the socket, just deep enough to allow its edge to lie in the same plane as the surface of the sole. The wedge is then put in, and secured by a couple of light blows from the hammer. The plane is then taken in the left hand, with the thumb resting on the wedge in the socket. The sole is turned upwards, and the iron is carefully driven in a little more, so that its edge shows just as much beyond the plane of the sole as the occasion requires. If it seems crooked, *i.e.*, if one corner seems lower than the other, this must be rectified by light taps on its free edges. When its position appears to be right, the iron is secured by driving the wedge in more firmly. If, after this, the iron is found to be too low, it may be made to recede by a blow on the back part of the stock, or, in the case of the trying-plane, by a blow on the *boss*, a piece of hard wood or metal inserted in front of the socket (see *H*, Fig. 54). [This *boss* is not always found in English planes. It is useful in slöjd as indicating the place to which the blow should be directed, and thus saving the stock of the plane from injury.—Trs.] The loosened wedge is then fastened once more, and the position of the iron is tested by the thickness of the shavings it removes, and raised or lowered, if necessary, according to the above directions. When the iron is removed, the plane is held in the way indicated above.

1. Planes with Flat Soles for the dressing of plane surfaces.

1. The **jack-plane** (Fig. 57). To give certainty and ease in working, the front portion of the stock of a Swedish jack-plane is furnished with a horn for the hand, and a metal

support of American invention is sometimes placed behind the iron to prevent the other hand from coming in contact with its sharp edges. The iron is single, *i.e.*, it has no cover, and the edge is curved, not square. The Swedish jack-plane is 9½ inches long. [The English jack-plane is 16 inches long.—TRS.] *Length of the jack-plane.*

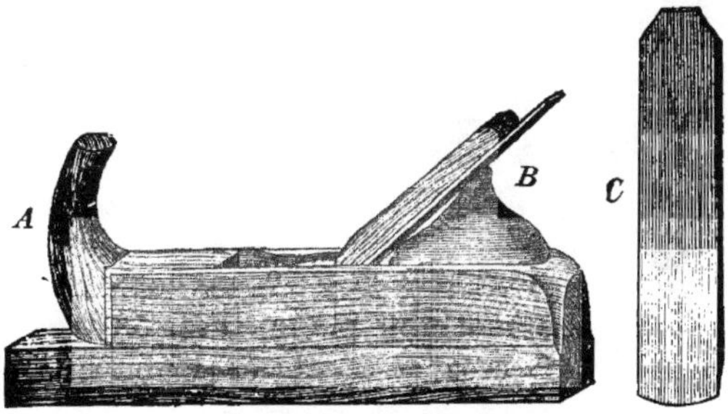

Fig. 57. Jack-plane. ¼.
A horn, *B* support for hand, *C* single iron.

The jack-plane is used on rough unplaned surfaces as a preparation for a finer plane, when the object in view is more to remove thick shavings rapidly by an iron which cuts deep, than to produce a smooth surface. As the iron is single, and the opening in front of it tolerably wide, the jack-plane has a tendency to tear up the wood; and it is therefore not advisable to use this tool very near the surface which is ultimately to be produced.

2. The **trying-plane** is the largest and most indispensable of all the planes in use. That it may be wielded steadily it is provided with a *handle* for one hand. The iron is *double, i.e.*, provided with a cover. Its various parts and their construction are fully described in connection with Fig. 54, and the method of using it is described in Chap. V.

It is employed in *shooting, i.e.*, in producing level surfaces of all kinds, and it is sometimes used in preparatory work instead of the jack-plane, in which case the iron should be set rather deeper than for shooting. When the trying-plane is used instead of the jack-plane, the space between the socket and the edge of the iron in front should be wider than in the later stages of planing. *Use of the trying plane*

Grinding the plane-iron.

In all planes used for shooting, the surface of the sole must lie altogether in the same plane ; and the edge of the plane-iron must be ground quite straight, and at right angles with the middle line of the iron. As, however, the corners of a perfectly straight-edge are apt to tear up the fibres by the side of the iron, or at least to leave a mark on the wood, they should be very slightly rounded. The sole is sometimes rubbed with raw linseed oil, that it may glide more smoothly over the wood.

The trying-plane should always be worked in the direction of its length, not obliquely to it, as is often improperly done.

Length of the trying plane.

The trying-plane should be about 20 inches long. [The English trying-plane is 22 inches long.—TRS.]

3. The **smoothing-plane** (Fig. 58) resembles the jack-plane, but is broader, and has a double iron.

Use of the smoothing plane.

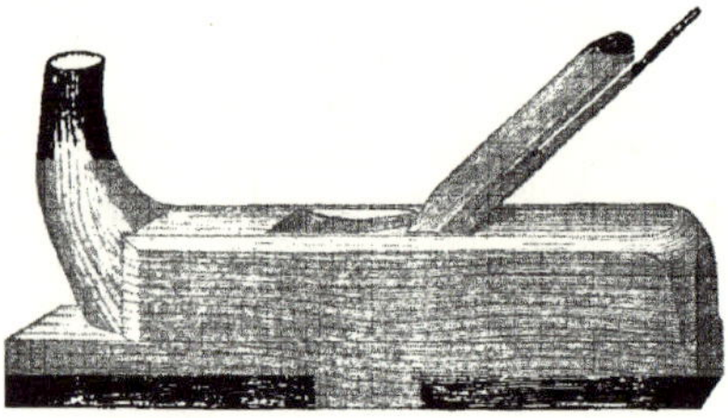

Fig. 58. Smoothing-plane. ¼.

The smoothing-plane is used after the trying-plane to produce a very smooth polished surface. As the shavings it removes must be extremely fine, the edge of the cover is placed very close to the edge of the iron, or, as it is called, is "set fine in front."

The smoothing plane should be about 9½ inches long. [The English smoothing-plane is 7½ inches long.—TRS.] The smoothing-plane and planes like it may be furnished with a support for the hand, behind the iron, like the jack-plane.

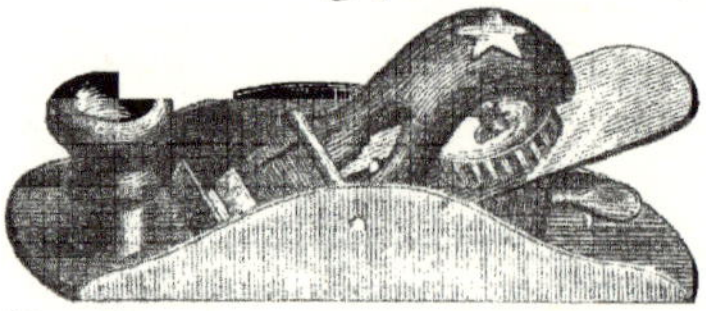

Fig. 59. Iron Smoothing-plane (American pattern). ⅕.

Iron planes

As mentioned above, the stocks of planes are sometimes

made of iron. Planes of this kind are used in England, and to a still greater extent in America. The plane-iron is adjusted by means of a screw. Small iron smoothing-planes are very useful for children, whose hands are not large enough to hold planes of the ordinary size. A plane of this pattern is shown in Fig. 59.

4. The **rebate-plane** (Fig. 60). When the adjacent surfaces of a rebate have to be planed, the ordinary smoothing-plane does not answer because the iron is narrower than the sole. In the rebate-plane the edge of the iron is as broad as the sole, sometimes even a little broader. The upper part of the iron is much narrower, and it is wedged into a mortise in the stock. The iron is single, and the shavings escape through an opening above its edge.

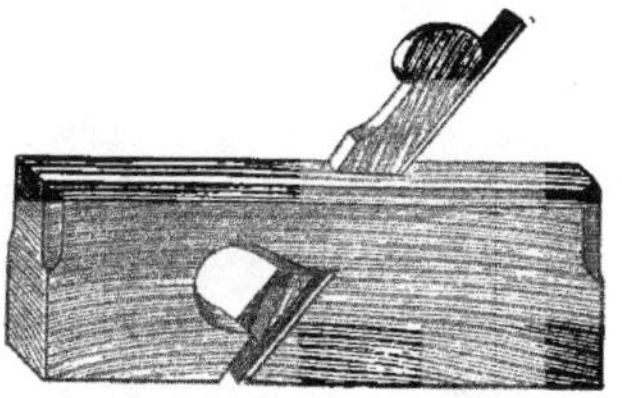

Fig. 60. Rebate-plane. ⅛.

2. Planes for the Dressing of Curved Surfaces.

1. The **round**.—This plane is used for hollow grooved surfaces. It resembles the smoothing-plane and the jack-plane, but differs from them in the more or less convex sole,—the degree of convexity depending on the degree of concavity it is desired to produce. The iron may be single or double, and the edge is rounded to correspond with the sole. An ordinary jack-plane may easily be converted into a round, by rounding the sole and the edge of the iron. In working the round

Fig. 61. Round. ⅛.
P seen from behind.

it must always be driven forward in a line with its length. In consequence of the shape of the tool, any other method would destroy the surface required.

[2. The **hollow**, another plane of this kind, has the sole concave, and an iron to correspond. It is used in planing round surfaces.—TRS.]

3. The **compass plane.**—In this plane the sole is curved lengthwise, and the iron is an ordinary double one with a straight edge. It is used in planing hollow curved surfaces. Soles of different degrees of curvature are required, according to the radii of the surface to be planed, but it is not necessary that the two should accurately correspond. The curvature of the sole must not be *less* than the curvature of the surface of the work, but it may be greater. The difference, however, if any, must be slight, because the two opposing surfaces must correspond closely enough to permit of the steady guidance of the tool. One compass-plane, therefore, will not suffice for surfaces of greatly varying curvature.

Fig. 62. Compass Plane. 1/5.

American compass-planes of iron, called adjustable planes, have flexible steel soles, which can be adapted to surfaces of different degrees of curvature. One plane of this kind is therefore enough.

3. The Old Woman's Tooth-Plane, and Dove-tail Filletster.

The **old woman's tooth-plane** is quite unlike the planes hitherto described. It consists of a block of wood on the inner side of which is fastened an iron, secured by a thumb-screw. (Fig. 63). The construction of some planes of this kind is much simpler; they consist merely of a parallelopiped piece of wood, in the middle of which is wedged a straight or curved iron. In this case the blade of a firmer chisel is often used.

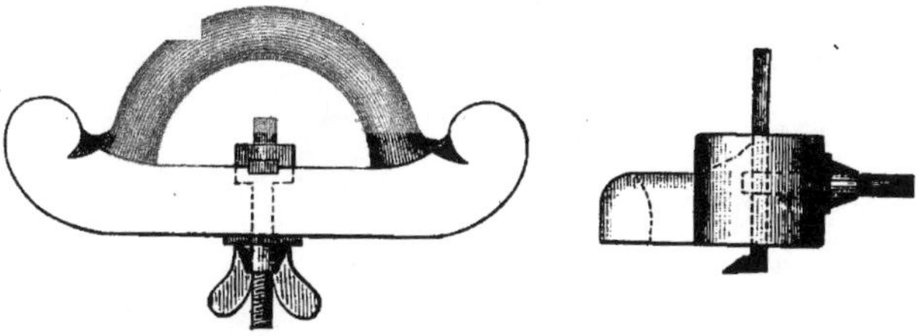

Fig. 63. Old Woman's Tooth-Plane, seen from above and from the side. ⅛.

The **dove-tail filletster** is like the rebate plane, but differs from it in having the plane of the sole oblique to the sides of the stock, instead of at right angles to them, and also in having a rebate either in a piece with the sole, or attached to it for the purpose of guiding the plane along the line of the dove-tail rebate to be formed.

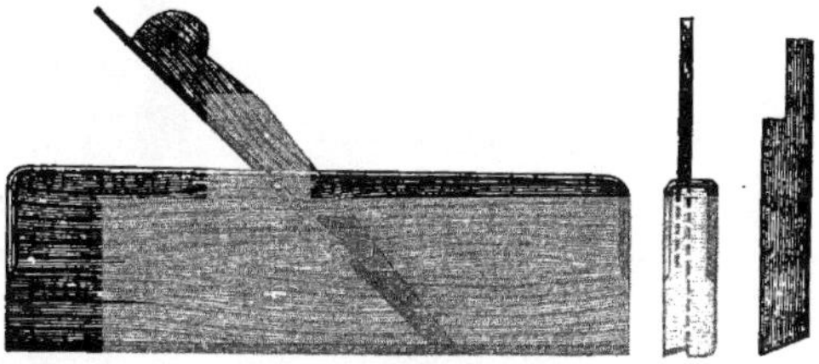

Fig. 64. Dove-tail Filletster, seen from the side and from behind. ⅛.

In the simple kind of filletster shown in Fig. 64, the rebate is fixed, but in the more complicated kind (Fig. 65), the rebate is adjustable to suit deeper or shallower work ; the latter is also provided with a "*cutter,*" which determines the line within which the surface is to be planed. This line, in other cases, must first be gauged with a cutting gauge; otherwise the plane will tear the fibres on that side and make it uneven.

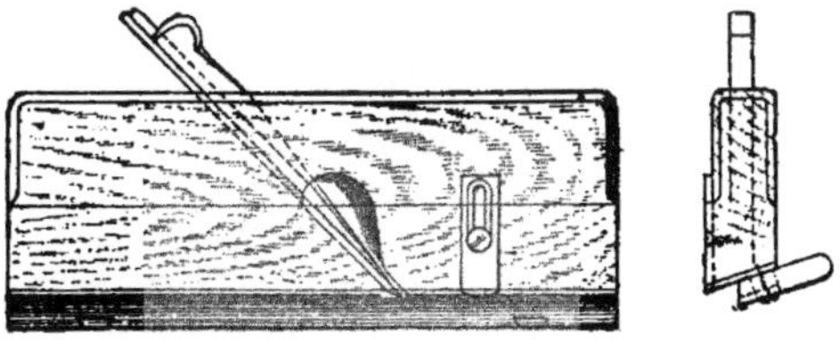

Fig. 65. Dove-tail Filletster. ⅛.

Both these planes are used in making dove-tail rebates; the old woman's tooth in smoothing and levelling the bottom of the groove into which the dove-tail is shot, and the filletster in working the dove-tail.

4. The Plough.

When a rectangular groove is made in a piece of wood the **plough** is used (see Plate X.) The breadth of the iron must not exceed the breadth of the groove to be made, and the sole consists of an iron splint set into the stock. The plough is furnished with a directing gauge, adjustable by bolts and wedges or screws. From 6 to 12 irons of different breadths accompany each plane.

5. The Iron Spokeshave.

The **spokeshave** may be included in the same class as the plane. It is made entirely of iron, with two handles, and is worked with both hands. The sole is very short—shorter than the breadth of the iron—and this renders the tool very useful in forming narrow convex or concave surfaces.

Fig. 66. Spokeshave. $\frac{1}{4}$.

The iron is secured by a screw and a fixing plate. The latter also does duty as a cover, and makes the tool more serviceable (see Fig. 66).

The spokeshave is a simple, practical, and easily-managed tool. It is made in several sizes, and the iron may have a *straight edge*, or one which *curves outwards*. The former is more common.

[The spokeshave described above is an American pattern. English spokeshaves are made of wood, and are recommended. —Trs.]

VII. Files.

The files used in wood-slöjd are the same as those used in metal work. The file plays, however, a much less important part in the former than in the latter. In wood-slöjd it is used chiefly to smooth curved surfaces, the interior of holes and depressions, and the ends of pieces of wood, in all cases where edge-tools cannot be used advantageously.

The **file** consists of a piece of steel, the shape of which may vary, and on the surface of which sharp *ridges* have been *cut* with a chisel. These ridges are equidistant the one from the other, and oblique to the length of the file. They form the *file-grade* (Fig. 67)—the essential characteristic of the tool. *The file-grade.*

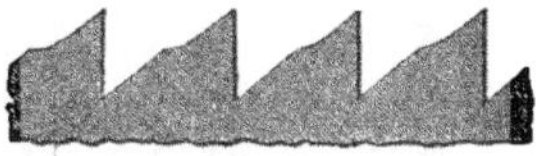

Fig. 67. File-grade. $\frac{8}{1}$.

A *single-cut* file is cut in one direction only; in a *double-cut* file the cuts cross one another. Both cuts incline towards the point of the tool, the result of which is that the file acts chiefly when driven forward, and has little effect when drawn back. The files used in wood-slöjd have usually a tapering point. All files terminate at the other end in a tang which slots into the handle.

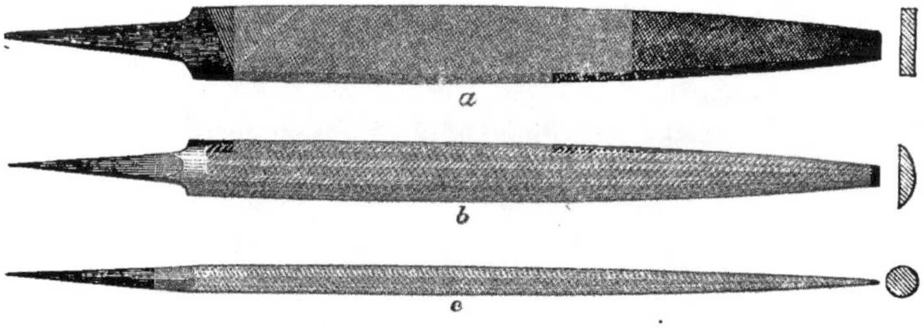

Fig. 68. Files. $\frac{1}{4}$.
a flat file, *b* half-round file, *c* round file.

Files are called triangular, square, flat, round, half-round, etc., according to the form of the blade in cross-section. *Flat round*, and *half-round* files are most used in wood-slöjd (Fig. 68). The *triangular* file is used for sharpening saws (Fig. 36).

The fineness of the file depends on the number of cuts per inch. They are usually classified as *coarse, medium, fine,* and *very fine.*

Medium files, about 12 inches long, are the most useful for working in wood, but coarse files, or *rasps,* may be used in the first stages of work.

Method of using the file.

When in use, the file is grasped by the handle by one hand, and the wrist or fingers of the other are laid on the point to produce the required pressure. The file is passed steadily and slowly backwards and forwards over the work if the surface desired is level, and with a circular motion if it is curved. *Pressure* is exerted only when the file is driven forward; when it is drawn back again it is allowed to glide over the surface. When the work cannot be made fast, the file must be worked with one hand; but, whenever possible, the work should be secured to the bench, that both hands may be free to direct and steady the file.

Cleaning the file.

In filing resinous or unseasoned wood, the cuts of the file are apt to become clogged with sawdust. The file may be cleaned with a stiff steel brush, but the simplest method of cleaning a wood-file is to wash it in hot water. The same file should never be used for wood and metal.

VIII. Methods of Finishing Work.

1. The Scraper.

Sharpening the scraper.

This tool consists of a highly-tempered piece of steel (Fig. 69). The edges of the scraper are generally straight, but sometimes the ends are rounded or hollowed to suit concave or convex surfaces. The two longest parallel edges are ground at right angles to the sides. When the

Fig. 69. Scraper. ¼.

scraper is sharpened, it is placed at the edge of a plank, and a very hard piece of steel is drawn against its edge as nearly as possible on the plane of the plank. This, when repeated several times backwards and forwards, levels the sharp edge of the scraper, which is raised up again by having the steel *once*, steadily but not too heavily, passed along it. During this the steel is held almost perpendicular, with its upper end inclined very slightly towards the upper side of the plank. The *raised* edge of the scraper now forms a fine edge, which takes hold of the wood when drawn across its surface, and removes minute shavings. When it becomes blunt, it must be sharpened once more, and as its edge, after repeated sharpening, becomes uneven, it must finally be re-ground. A worn saw-file, the cut of which has been carefully ground off, and the edges slightly rounded, or a firmer chisel, may be used.

Method of using the scraper.

The scraper should be held easily in the hand. In polishing a plane surface, the tool should be taken in both hands. The scraper should incline towards the surface of the work (see Chap. V., page 136), and should be worked always in the direction towards which it leans, and *with* the grain of the wood, but somewhat obliquely to the direction of the fibres. Towards the end, pressure should be diminished, to produce a finer polish. Care must be taken lest the cutting edge become ragged from careless "setting," and scratch the surface. Should this be so, the scraper must be re-ground and then sharpened.

2. Sand-Paper.

Sand-paper made of flint and of glass.

Sand-paper is made of paper with a coating of finely-ground flint, glass, or quartz glued on to it. The grains on the same paper are always of the same size, and, according to the finer or coarser quality, the paper is numbered from 0 to rough 2.

[Sand-paper made of flint is generally used in Sweden. In England, glass-paper is considered the best.—Trs.]

When in use, the sand-paper should be torn off in pieces of convenient size, and a bit laid on the plane surface of a piece of cork or wood, ¾-inch thick, and of a good size to be held in the hands. If a sufficiently thick piece of cork cannot be obtained, a thin piece should be glued on a piece of wood, or, failing cork, a piece of card-board will answer the purpose. This serves the purpose of a soft *rubber* (wood alone being too hard), and gives the necessary support to the sand-paper, which, used in this way, acts much in the same manner as the file, and may be considered to all intents and purposes as a tool.

Sand-paper really a tool.

Sand-paper may be used without a rubber only in the case of concave or convex surfaces, where there are no sharp edges. Care must be taken in finishing off not to work the paper in the direction of the fibres, but either at right angles or obliquely to it, in order to produce a smooth surface. Just at the last, the paper may be passed once or twice in the same line as the fibres, to remove any ridges or marks which may have been produced. For similar reasons, the paper last used should be finer than that first employed, in order to secure a perfectly smooth surface.

Sand-paper should be used sparingly.

Sand-paper should never be used to form or smooth up the surface of objects. The knife, the file, the smoothing-plane, the scraper, etc. are the proper tools for this purpose. Sand-paper should be used only in finishing off, and when the use of the smoothing-plane is understood, it is not much needed for plane surfaces. In the case of objects with curved surfaces, on the other hand, it is almost indispensable.

Finishing off with sand-paper should never be done in a thoughtless, mechanical way. To attain a satisfactory result the greatest attention is requisite.

IX. Brace and Bits.

Bits of different kinds are used in making round holes. Bits for wood are made of a special kind of steel, one end of which forms the cutting portion of the tool, and the other is wedge-shaped, that it may be securely fastened into a handle or *brace*, by means of which it revolves. In working, the brace is always turned *to the right*, and the bits are made to cut in the same direction. The edge of the bit is designed to make its way into the wood without great pressure, and without risk of splitting it. The bit must work without hindrance from the shavings; otherwise it will become hot from friction, and boring will be difficult. A good bit cuts like a knife, detaches smooth spiral shavings, and becomes only moderately warm, even when worked quickly.

How a bit should work.

The **brace** may vary in construction. Fig. 70 shows a very strong Swedish brace made of iron. The upper end, or *tang* of the bit, forms a square truncated pyramid, which slots into a hole in the *brace socket*, and is fastened by a spring. Fig. 71 shows an American brace, also made of iron. It has a *screw adjustable socket*, into which the bit is secured. The tang of the bit may be of any form, provided it is somewhat rectangular.

Fig. 70. Swedish Brace. $\frac{1}{4}$.

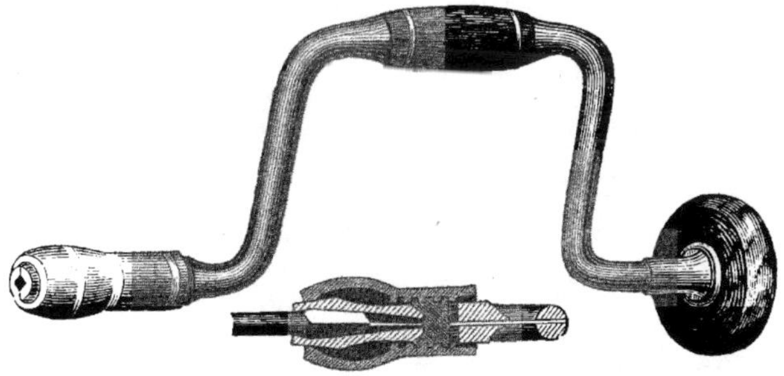

Fig. 71. American Brace; section of screw adjustable socket, or bit holder. $\frac{1}{4}$.

The bits in most general use are *shell-bits* and ***centre-bits.*** Small shell-bits are called pin-bits.

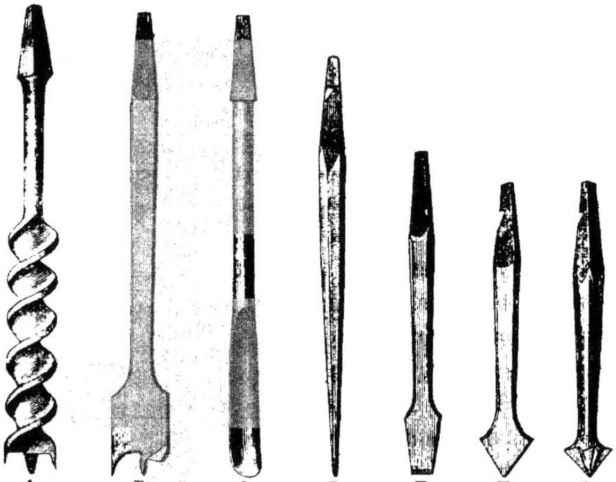

Fig. 72. *A*, Auger-bit. *B*, Centre-bit. *C*, Shell-bit or pin-bit. *D*, Hole-rimer drill. *E*, Screwdriver bit. *F* and *G*, Counter-sink drills. $\frac{1}{3}$.

1. The **shell-bit** (Fig. 72, *C*) is gouge-shaped, with the end curved like the point of a spoon. Unlike the centre-bit, it has no middle point, and it is therefore more difficult to gauge to holes of any given size, especially if the latter are large.

This bit is better than the centre-bit for boring end pieces. Shell-bits are made in various sizes, from those adapted for holes of $\frac{1}{16}$ inch in diameter to those suitable for holes of $1\frac{1}{2}$ inch in diameter. The smallest kind, the pin-bit, is most used in wood slöjd. A set of pin-bits includes from 8 to 10, varying in size from $\frac{1}{8}$ inch to $\frac{1}{2}$ inch.

2. The Centre-bit.

The ordinary centre-bit.

1. The ordinary centre-bit (Fig. 72, *B*, and 73, *B*) has a flat blade, the lower portion of which is broader than the upper, as is shown in the illustrations. In the middle of the lower edge is the *centre-point a*, and at one side is the *cutter b*. In

boring, the cutter makes a circular incision corresponding to the circumference of the hole, and thus determines its diameter, prevents the wood from splitting, and facilitates the removal of shavings and sawdust by the *lip c*, the edge of which is horizontal to the point and oblique to the blade, and which cuts at right angles to the cutter. The centre point is longer than the cutter, which again cuts deeper than the lip. In sharpening the centre-bit, which may be done with a small half-round file, care must be taken that the edge of the cutter is on the outer side, and the edge of the lip on the under side. A set of bits should contain 8 to 12 centre-bits, from $\frac{3}{8}$ inch to 1 inch broad.

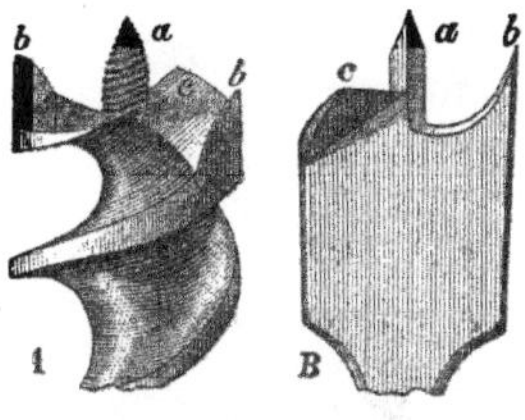

Fig. 73. *A*, Portion of Auger-bit. *B*, Portion of Centre-bit. *aa* Centre-point, *bbb* Cutter, *ccc* Lip.

2. The **auger** (Fig. 72, *A*) belongs to the same class. In boring with the bits previously described, it is necessary to exercise a certain degree of pressure, but the auger works its way into the wood by means of the conical screw which forms its centre point, and after the screw has once started all that is needed is to make it revolve. The auger is besides furnished with a cutter and a lip on both sides of the screw. (See Fig. 73, *A*.) *Way in which the auger works.*

Above the cutting portion it is spiral in form, and thus we have a double spiral with sharp edges. This gives plenty of room for the sawdust and shavings which are worked out of the hole without the removal of the auger. The American augers are the best. A set includes 6 to 12 pieces, from $\frac{3}{16}$ inch to 1 inch. *The best augers*

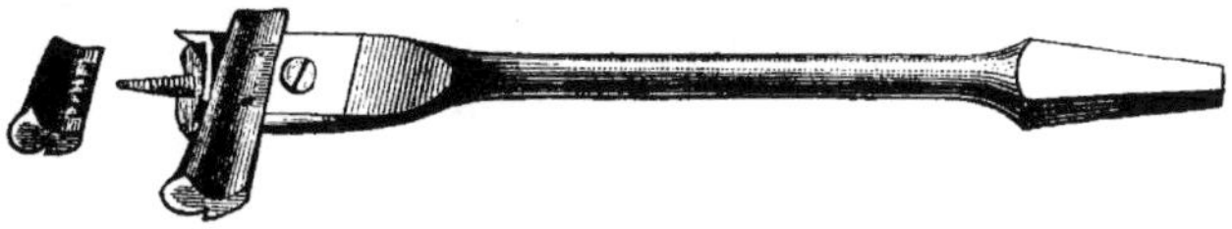

Fig. 74.—Expansion bit. $\frac{1}{3}$.

Adjustable or expansion bits.

The expansion bit (Fig. 74) is of American construction. Within certain limits it admits of holes of different sizes being bored with one and the same bit. Its point, lip, and cutter are tolerably like those of the auger, but it is furnished with two loose cutters, which may be screwed in to suit the diameter desired. The adjustable cutter does the work both of lip and cutter. The expansion bit makes holes with remarkably even surfaces, and with two different sizes it is possible to bore holes varying in diameter from ¼ inch to 3 inches.

Screwdriver bit, counter-sink drill, and hole-rimer.

A screw-driver bit (Fig. 72, E), two or three *counter-sink drills* (Fig. 72, F and G), and a couple of square or hexagonal *hole-rimers* (Fig. 72, D) are usually included in a complete set of bits. The counter-sink drill is used to produce a conical hole in wood or metal, suitable for sinking screw-heads. The rimer enlarges holes in thin metal plates, *e.g.*, screw holes in hinge-plates.

Fig. 75. Bradawl. ⅓.

3. **The Bradawl** (Fig. 75). This tool consists of a steel bit $\frac{1}{16}$ inch to ⅛ inch thick, and 2 inches long. Its point is like that of an awl, or it may be chisel-pointed. The bit is secured in the handle by a screw-socket. Several bits of different sizes belong to the tool, and the handle, which is hollow, serves as a case for them. The bradawl is used to bore holes for sprigs, nails, etc. When holes are bored with the chisel-pointed bit, the edge is placed across the grain of the wood, and pressure is exerted in this direction to prevent the splitting of the wood.

X. The Mallet, the Hammer, the Hand Vice, Pincers, and Screwdriver.

The Mallet (Fig. 76) is made of hard, strong wood, preferably of figured beech. It is used for striking tools with wooden handles, because the hard hammer in such cases would not only do damage, but would not serve the purpose so well.

The Hammer (Fig. 77) consists of a piece of steel with a hole for the handle, called the *eye.* One end is cylindrical and terminates in a flat surface, called the *face;* the other end, which is called the *pane,* is wedge-shaped, with a rounded edge. That the handle may be quite firm, the eye widens at the sides, and wedges driven hard into the end of the handle cause it to fill up the cavity entirely.

Fig. 76. Mallet. ⅛.

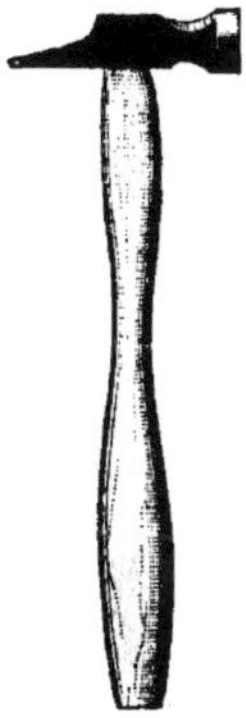

Fig. 77. Hammer. ⅛.

There are various kinds of **pincers,** but only those used in wood slöjd need be named here. Pincers have two steel arms rivetted together. The rivet divides the arms into two unequal portions, the longer, or *handles,* and the shorter, or *jaws.*

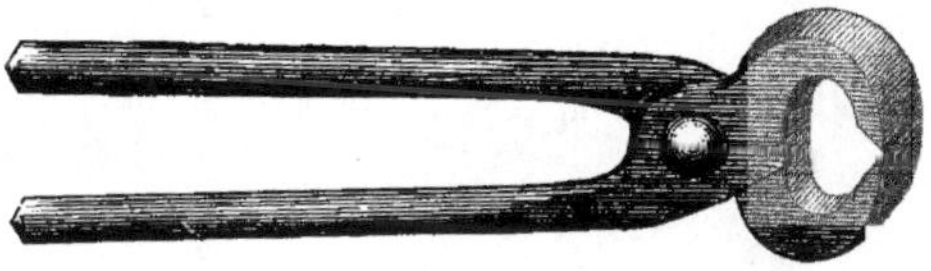

Fig. 78. Pincers. ¼.

The ordinary **pincers** have short, broad, sharply curved jaws, and are used to extract nails, etc.

The **wire-cutter** resembles the preceding, but is slighter in make, and its arms are curved and its jaws sharper. It is used to snap off pieces of wire, tintacks, etc. The jaws of the **flat pliers** are flat on the inner

Fig. 79. Wire-cutter. ⅓.

Fig. 80. Flat-jawed Pliers. $\frac{1}{3}$.

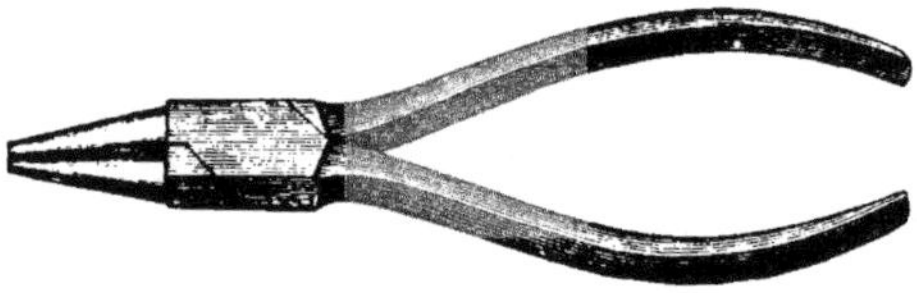

Fig. 81. Round-jawed Pliers. $\frac{1}{3}$.

side, which is file-cut to enable them to take fast hold of small pieces of metal to be filed, bent, &c.

In the **round pliers** the jaws are more or less conical in shape, for the bending of wire, etc.

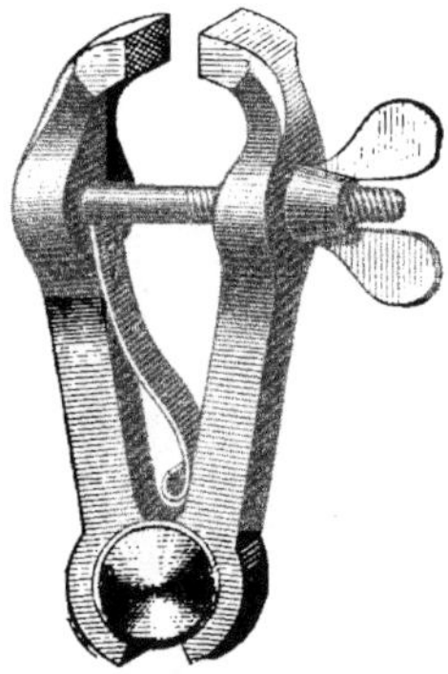

Fig. 82. Hand-vice. $\frac{1}{3}$.

The **hand-vice** is not so much employed in slöjd-carpentry as in metal slöjd. Its chief use is to secure small pieces of metal for filing. It may be held in the hand, or, after the piece of metal has been made fast, it may itself be screwed into a hand-screw or to the bench, that both hands may be free for the work of filing.

a

Fig. 83. Screwdriver. $\frac{1}{3}$

a the point, full size, seen from the side.

Shape of the screwdriver.

The screwdriver is used for driving in screws, and is made of hard steel. At the end it is bevelled to a thick point, which varies from $\frac{1}{64}$-inch to $\frac{1}{16}$-inch in thickness, depending on the size of the screw for which it is to be used.

The bevelled edges should be parallel, and the point should be as little as possible like a wedge in shape, but should lie flat in the slit of the nail, otherwise it will have a tendency to slip and become chipped.

E. The Grinding and Sharpening of Tools.

Work must never be done with blunt or badly-set tools. Tools must always be kept sharp and in good order.

These rules should *always* be kept in mind. Many a slöjder toils in the sweat of his brow with a blunt saw, or a badly set blunt plane, rather than take time to put his tool in order, though tools in good condition save hours of work, much unnecessary trouble, and needless vexation. Blunt tools demand more strength and exertion than sharp ones, and seldom, if ever, produce such good results. The rules given above are especially important in the case of children, for whom work ought not to be made unnecessarily difficult.

The sharpening of edge-tools is performed on the *grindstone* and the *oilstone*. The method of sharpening a saw has already been described (pp. 80, 81).

The ordinary **Grindstone** consists of a circular slab of sandstone, which rotates on an axle, and is provided with a handle for turning. It is supported on a grindstone stand or bench. Below the stone is a wooden well, lined with zinc, partially filled with water, into which the stone is sunk about one inch when in use. The stone should not be too fine in the grain or too hard.

Care of the grindstone.

The grindstone should never be used dry, because the steel does not "catch" well unless the stone is wet, and the friction on a dry stone "burns" the steel and makes the edge of the tool soft. Exposure to the sun for any length of time makes the stone too hard, while prolonged immersion of any portion of it in water renders that portion soft. Consequently it wears faster, and the stone becomes uneven or *eccentric*. The stone should therefore be kept dry except when in use.

A frame-work attached to the stand prevents splashing when the stone rotates by directing the water down into the well, and splashing may still further be avoided by fastening a thick piece of stuff in front so that it trails upon the stone and absorbs a portion of the surplus water. The stone must always be turned *towards* the worker and towards the edge of the tool, which must be moved steadily, and with equal pressure from side to side, across the whole breadth of the stone, to prevent the formation of scratches or depressions on its circumference. The bevelled edge produced by grinding must present either a flat or a concave surface to the convex surface of the stone. It must never be convex. The concave form of the bevelled edge is advantageous, because it materially lightens the final sharpening on the oilstone. The edge must also be quite straight unless a curved edge is actually required.

Straight edge.

As it is difficult, especially for the inexperienced, to hold the steel steadily enough against the stone, a grinding support has been invented. Such a support of American make is shown in Fig. 84. It consists of an iron frame into which the plane-iron or the chisel is screwed. A small wheel below

Accurate grinding.

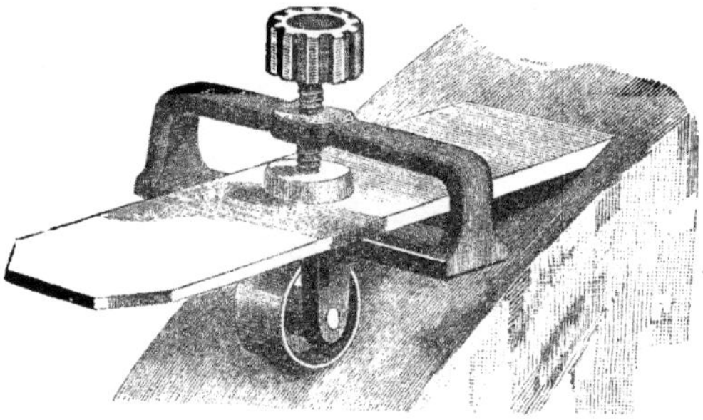

Fig. 84. Grinding support. ⅓.

the frame revolves upon the grindstone, and the desired angle on the edge of the tool is obtained by fastening it in with the edge at a shorter or longer distance from the frame. By means of this simple contrivance even an inexpert pupil is able to grind a plane-iron correctly.

A very common fault in grinding is to make the angle which the bevelled edge makes with the face of the tool too great, *i.e.*, to make the edge too thick. This is often done by beginners in their haste to be relieved from grinding.

The tool must be ground till a *raw edge* appears, *i.e.*, the very thin "film" or hair produced by the grindstone's removing the very edge of the steel. This, in its turn, is removed by the oilstone. *The raw edge.*

Sharpening with the oilstone is necessary, because the edge produced by the coarse-grained grindstone is neither fine enough nor even enough for immediate use.

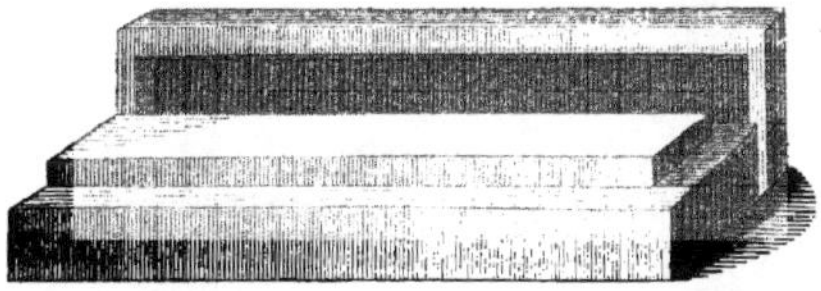

Fig. 85. Oilstone and case. ¼.

The **oilstone** is a slab of specially fine-grained stone. "Washita" and "Arkansas" stones from America, and "Turkey" stones are the best. [Welsh oilstones are less expensive, and can thoroughly be recommended.—Trs.] The oilstone should be 8 inches long and 2½ inches broad, and it should be kept in a wooden box with a cover (Fig. 85). A good oilstone is very hard and close-grained, and it "takes well," *i.e.*, it acts almost like a very fine file on the steel. The colour is yellowish-white. They last a long time, but are expensive to buy. *The best oilstones.*

When in use, the oilstone should be moistened with vegetable oil. The addition of a little paraffin is an improvement. The tool is held in both hands, and the bevelled edge is applied closely to the stone in such a way that, while the bevel is altogether in contact with the stone, the edge presses rather more heavily on it, and this angle of inclination must be steadily maintained to prevent the edge from becoming rounded. The steel is now drawn over the stone with a slow, steady, backward and forward motion. When this has been repeated often enough, it is turned over and passed once over the stone with the face flat. The worker must not confine his operations to the middle of the stone, but must use the whole of the surface. *Method of using the oilstone.*

An *oilstone slip, i.e.*, a piece of the same kind of stone as the oilstone, but smaller and thinner, and rounded at the edges, is required for the sharpening of gouges, spoon-irons, etc.

A sharp edge. Sharpening must be continued until the edge itself is not visible when held up against the light, or until it no longer appears white and rounded. Its sharpness is tested by touch ing it lightly with the finger.

F. The Tool Cupboard.

For the benefit of those who wish to procure a tool cupboard, complete drawings of one are given in Plate XI. It is so arranged that every tool has a fixed, easily observed place, in order that the absence of any may be readily discovered when the tools are laid past. Tools must further be so arranged that when *one* is taken out *another* is not displaced; and all sharp edges must be protected.

Any alterations in the size of the cupboard, required by a larger or smaller stock of tools, could easily be made.

CHAPTER IV.

JOINTING.

Different parts of articles are connected or *jointed* partly by *glue, nails,* or *screws,* and partly by the special adaption of the parts themselves, as in *mortising* and *dove-tailing.*

A. Glueing.

The simplest way of jointing two pieces of wood is to introduce between them a connecting medium in liquid form, *i.e., glue.*

Glue is made from the refuse, clippings, etc., of tanneries and glove manufactories. After being subjected to a boiling process, these materials are reduced to a viscous fluid, which solidifies on cooling into a stiffish jelly, which is then cut into thin slices and dried upon nets stretched on frames.

Good glue is known by its light brown or brownish yellow colour; its sparkling transparency; its hardness and elasticity; by the way it breaks off in flakes and whitens in the line of fracture; and by its power of resistance to the dampness of the air. It swells if steeped in cold water, but does not melt even after one or two days' immersion. The ultimate test of good glue is, however, its cementing power.

1. The Preparation of Glue.

The cakes of glue, either entire or in pieces, are first soaked in cold water. After the glue swells it is put in a glue-pot (Fig. 86) and melted by heat. The glue-pot consists of two pans usually made of cast-iron or tin-plate. The larger of these, the *outside pan,* is, when in use, half filled with water, and the smaller one, the *inside pan* or *glue-pot* proper, in which the glue is placed, rests upon a rim or *flange* round its mouth. This inner pan should always be lined with tin. The water in the outer pan prevents the glue from burning, (an accident which must always be carefully avoided), and as the contents of the glue-pot are surrounded by warm water, they may be kept fluid and fit for use a considerable time after the pan has been removed from the fire.

If glue is wanted *in a hurry,* the cakes may be put in a towel or a similar piece of stuff to keep the glue from being scattered about, and broken to pieces with a hammer. The pieces are then put into the glue-pot and stirred during boiling, to prevent unmelted glue sticking to the bottom. This mode of preparation is quite as good as the preceding.

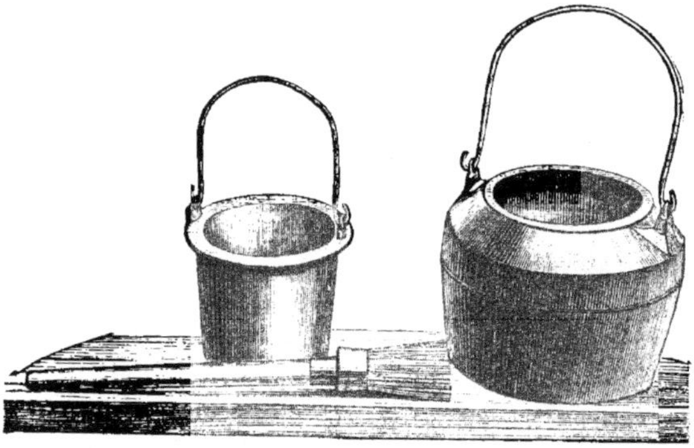

Fig. 86. Glue Pot (inside pan) and Brush. Outside Pan. ⅙.

Glue is applied with a strong brush, of which there should be two sizes, one for large surfaces and one for small surfaces, *e.g.*, mortise holes, etc.

Liquid Glue.—The addition of acetic acid to melted glue prevents putrefaction, and, without lessening its cementing power, keeps it liquid at ordinary temperatures. "*Liquid glue*" may be made as follows :—Four parts of good glue are melted in four parts diluted acetic acid, in the outer pan, or on the top of an oven. One part spirits of wine and a small quantity of alum are then added, and the mixture is kept in a wide-mouthed bottle, the cork of which has a hole to admit the brush.

This glue remains liquid at + 14° to 18° C., and does not solidify until + 8° to 12° C.; it is very convenient for small articles, as it is always ready and in good condition, and its cementing power is quite equal to that of glue prepared in the ordinary way. Its only drawback is that it dries more slowly.

In the case of articles exposed to moisture, the addition of 10 per cent. of boiled linseed oil is advantageous. The glue to which it is added should be hot and strong, and should be stirred till the varnish has been thoroughly mixed. The wood to which this *wood-cement* is applied should be dry and warm, and the pieces should be firmly pressed together until the glue dries.

2. Glueing.

The process of glueing is very simple, but it must be carefully performed to ensure a strong inconspicuous joint. The general rule holds good that the layer of glue shall be so thin that the seam can hardly be seen, and this presupposes that the pieces fit accurately (see page 146), that they are kept in sufficiently close contact while the glue is drying, and that the glue itself does not cool before they are put properly together.

Warming the wood. To keep the glue from cooling, the wood should be warmed as well as the glue, and the operations of applying the latter, putting the pieces together and applying the required pressure, must be rapidly performed. Generally speaking, it is sufficient if one of the wooden surfaces is warmed: thus in dovetailing and slotting the pins only are warmed; in blocking, the blocks only, etc.

Laying on the glue. The glue, which must be neither too thick nor too thin, is laid evenly and quickly, in as small a quantity as possible, over the surface of the wood with the brush.

In the case of pins for mortising, the glue should be thicker than for jointing boards, and the glue is generally applied to the hole as well as to the previously warmed pin, though sometimes only to the latter.

Screwing together is performed either in the bench, which is the simplest method, or in hand-screws, or in a press with wedges. The article must remain under pressure till the glue dries. If the glue is too thick or the wood cold, or if the glue cools before screwing up, the joint will show, and will not be good. A joint of this kind does not look well, and is less durable than one properly glued together.

Making a joint. The bench pegs or the hand-screw should always be in order before glueing, to save time. Just before the final tightening of the screw, the work should be carefully examined to see if the parts are *in their right places.* If not they must be made to fit. If the staves of a barrel are not in the same plane, the screw must not be loosened, but the stave which is not flush must be hammered into place, and the screws tightened. The work must not again be disturbed till the glue has hardened.

In screwing up finished pieces of work, bits of wood must *always* be put between the work and the bench-pegs or the point of the screw, to prevent marks. When large plane surfaces are glued together, it is necessary to use several cramps to obtain strong enough pressure.

Removal of superfluous glue. The glue which exudes from the joints of objects which are finished off before glueing, *e.g.*, the inside of a drawer, must be carefully wiped off with a clean sponge or rag dipped in warm water immediately after glueing together, before it completely dries. Care must be taken not to wet the wood unnecessarily.

The better the glue penetrates the pores of the wood, the stronger the joint. Consequently, glue holds better in loose-fibred than in close-grained wood, which presents a hard, smooth surface. Broad surfaces of the latter description are roughened a little before glueing, by drawing a coarse file over them.* Glue which dries slowly is stronger than that which dries quickly.

A well-fitting joint made with good glue is so strong that, when long boards are joined together, the wood itself generally gives way before the joint. This, however, is not the case when end pieces are joined together, or when the wood is very hard or close-grained. *Strong and weak glue joints.*

Two pieces of wood may be glued together without *cramping* or screwing together, *e.g.*, a block of wood on a plank. The block only is warmed, but glue is laid upon both. The former is then pressed upon the latter, and rubbed backwards and forwards to get rid of the superfluous glue, until it begins to adhere. Care must now be taken that it is in its right place, and is not further disturbed. The two pieces adhere by atmospheric pressure.

B. Nailing.

Sprigs of different lengths and thicknesses are generally used for nailing together slöjd-work, but for large or heavy articles *cut or beat nails* are employed, because their uneven surface is more tenacious, and thus gives greater strength to the joint. Before the nail is hammered in, a hole should be bored with the pin-bit or the bradawl to prevent splitting. The diameter of this hole should not exceed two-thirds of that of the thickest part of the nail, and the nail should be hit straight on the head, to prevent it from bending or going in crooked.

The *firm hold* of the nail in the wood depends partly on the more or less rough nature of its surface, partly on its length and thickness, partly on the kind of wood, and partly on the direction of the nail in relation to the fibres, *i.e.*, whether it is driven into a long board or into an end piece. The strongest joint is made with beat nails in a cross piece; *The strength of nail-joints*

---

* There is a special tool for this purpose used in veneering, &c., called the "*toothing-plane.*"—Trs.

the weakest with beat sprigs in an end piece. The preliminary boring does not affect the hold of the nail in the wood unless it is too deep or too wide. A hole half the depth and half the diameter of the thickest part of the nail exercises no noticeable influence on the strength of the joint.

Sinking the head of a nail.

Sometimes it is necessary to *sink the nail* under the plane of the surface, that it may not present any obstacle to smoothing up or finishing off the work. After the nail has been hammered in by the ordinary method, a small steel *punch* about 4 inches long and ¼ inch thick, tapering to a thick point rather less in diameter than the head of the nail, is used to sink it. The *punch* is placed on the head of the nail, and hammered till the head sinks to the depth required.

Wooden pins are sometimes used for jointing. They are made of straight, split wood, and have four sides, often with bevelled corners, tapering slightly to a blunt point. They are driven into holes previously bored which their bevelled corners enable them to fit closely without splitting the wood. Glue is often added to strengthen their hold. These wooden pins are called *dowels.*

C. Screwing together.

Jointing with wood screws.

Wood-screws, *i.e.,* metal screws with thin, deep, sharp-edged tap-worms, are used for screw-joints. Screws which are gimlet-pointed penetrate the wood more easily than others.

The **wood-screws** used in screw-joints are of different kinds, with half-round, curtailed conical, or square heads. The two first only are used in wood-slöjd. In both, the head of the nail is furnished with a slit for receiving the screw-driver.

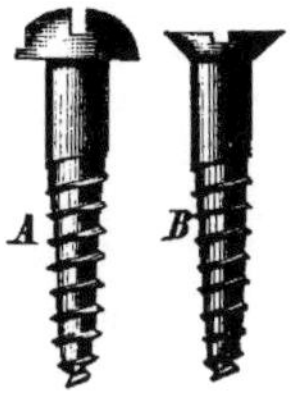

Fig. 87.
Wood-Screws.
A with half-round,
B with conical head.

When *A* (Fig. 87) is used, the head of the screw remains *above* the surface of the wood. In joints made with *B*, the head is made to lie *level with* the surface, for which purpose the hole bored for its reception is afterwards counter-sunk. Wood-screws are made in many lengths from about ¼ inch to 3 inches, and of varying thickness. They are very generally used, and are especially useful for articles which require sometimes to be taken apart and put together again.

Strength of a screw joint

In consequence of their peculiar form, screws give a much stronger and firmer joint than nails, which hold the pieces together simply by friction. A screw cannot be drawn out without unscrewing, unless the wood around it is cut away. The hole bored for the reception of the screw should be as deep as the length of the unwormed portion.

D. Jointing by means of the formation of the parts of the joint.

The names only of the various kinds of jointing of this nature are given below. A description will be found in Chapter V.

1. *Halving.*
2. *Mitreing.*
3. *Slotting.*
4. *Mortise* and *Tenon-jointing.*
5. *Groove-jointing.*
6. *Dove-tailing.*
7. *Hooping.*

E X E R

Number of Exercise.	Name of Exercise.	Purpose of Exercise.
1	*Long cut.*	To cut off a piece of wood in the direction of the length of the fibres.

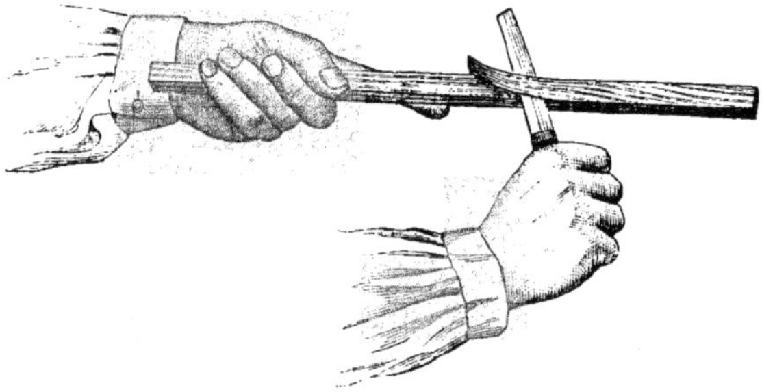

Fig. 88.

2	*Cross cut.*	To cut off a piece of wood at right angles to the fibres.

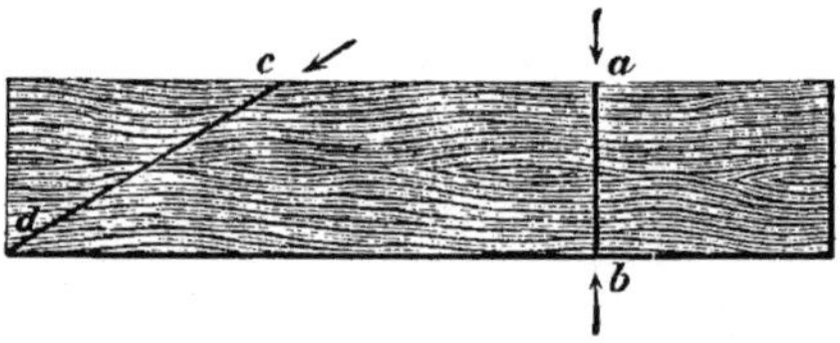

Fig. 89.

3	*Oblique cut.*	To cut off a piece of wood *obliquely* to the fibres.
4	*Bevel cut.*	To cut off a piece of wood in the direction of the length of the fibres in such a way as to produce a surface at an oblique angle to the adjacent surfaces.

Tools required.	Directions for Work.
Knife.	The knife is taken firmly by the handle, and the cut is made always *in the direction of the fibres*, but *away from the worker.* To steady and strengthen the hand which holds the wood, and to render the exercise easier, the piece of wood should always rest upon a board laid on the bench.
Knife.	The cut is made from both sides to avoid splitting (*a—b*). If the cut is short, the wood is laid upon the bench. If it is long, the wood is held in one hand and the upper arm is pressed against the body to secure greater strength and support during the exercise. For the manner of holding the knife see Ex. 1. For the proper position see Plate I.
Knife.	The cut is made in the direction of the fibres, not contrary to it. For the manner of execution see No. 2 (Fig. 89 *c—d*).
Knife.	For the manner of execution see No. 1.

Number of Exercise.	Name of Exercise.	Purpose of Exercise.
5	*Sawing off.*	To saw off a piece of wood at right angles to the fibres.
6	*Convex cut.*	To cut off a piece of wood, convex in shape.

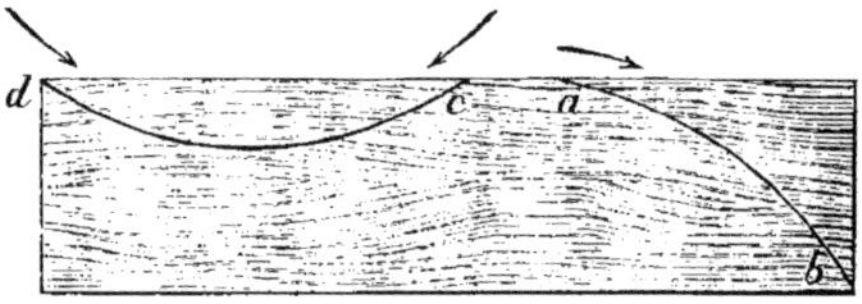

Fig. 90.

Number of Exercise.	Name of Exercise.	Purpose of Exercise.
7	*Long-sawing.*	To rip up a piece of wood lengthwise.
8	*Edge-planing.*	To plane a piece of wood, the surface of which is narrower than the plane-iron.

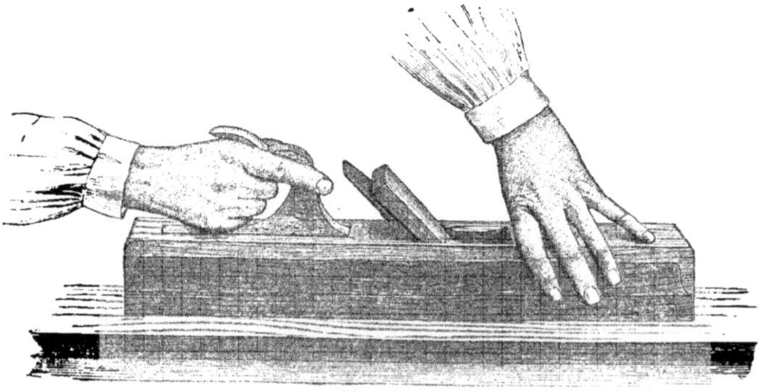

Fig. 91.

Tools required.	Directions for Work.
Broad-webbed bow-saw.	The piece of wood is screwed into the bench, and the saw is worked with long, gentle strokes parallel with the edge of the bench (see Fig. 89 *a—b*, and page 84). The final strokes must be made cautiously, because the wood may easily be split. Before beginning the exercise, the worker should see that both edges of the saw are in the same plane, and that the teeth of the saw point *away* from him.
Knife.	The fibres are cut obliquely (*a—b*). See further under No. 3. (See also Plate I. for the position of the worker.)
Broad-webbed bow-saw.	For the method of execution, see No. 5. See also Plate II. for the position of the worker.
Trying-plane.	The piece of wood is fastened between the bench-pegs so that it lies firmly and evenly upon the bench. Before the plane is used it should be carefully set for the particular kind of wood to be planed, *i.e.*, the plane-iron should come lower down in the case of loose-fibred wood than for hard wood, and the cover should be placed farther from the edge of the iron in the former case than in the latter (see page 98). The handle of the plane is firmly grasped in one hand, and the other is placed right in front of the socket. The plane is then worked briskly to and fro over the surface. The path of the plane must always be horizontal, regulated by the difference in the pressure given by one or other of the hands. For the position of the worker, see Plate III.

Number of Exercise.	Name of Exercise.	Purpose of Exercise.
9	*Squaring.*	To prove whether two plane surfaces in a piece of wood are at right angles.

Fig. 92.

Number of Exercise.	Name of Exercise.	Purpose of Exercise.
10	*Gauging.*	To produce parallel lines at a given distance from the edge of the work.

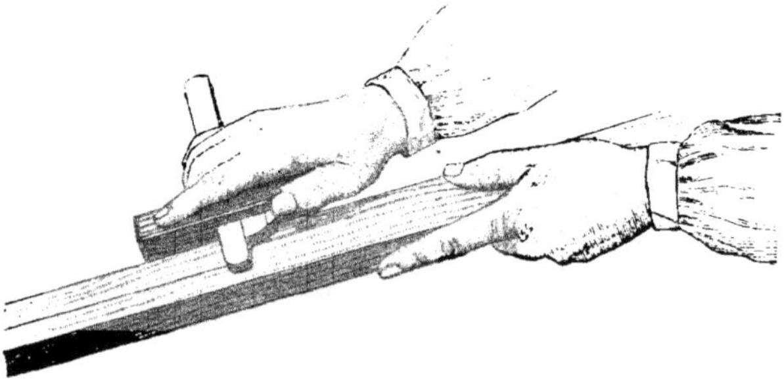

Fig. 93

Tools required.	Directions for Work.
Square.	The stock of the square is grasped in one hand; and its inner surface is applied close to the *face* of the work (*i.e.*, the side first planed), while the blade rests upon the other side.
Marking gauge.	The stock of the marking gauge is held steadily and closely to the faced-up sides of the work parallel to which the line is to be made.

Number of Exercise.	Name of Exercise.	Purpose of Exercise.
11	*Boring with the shell-bit (pin-bit).*	To make a hole of small diameter.
12	*Face-planing.*	To plane a piece of wood when the surface is broader than the plane-iron
13	*Filing.*	To dress up rough surfaces.
14	*Boring with the centre-bit.*	To make a hole of large diameter

Tools required.	Directions for Work.
Shell-bit (pin-bit).	The object in view is partly to make a hole and partly to avoid splitting the wood, when sprigs, larger nails, or screws are put in. The left hand is laid upon the brace stock, to give pressure from above downwards; the right hand grasps the handle in the middle, and the brace is turned towards the right, care being taken that the centre of the bit enters the right place in the wood, and that the direction of the hole is perpendicular to the plane of the work or the bench. The latter condition presents some difficulty, especially to beginners, and is best fulfilled by the slöjder's standing alternately in one of two positions, in order that he may see the bit from each side of a right angle. To give greater pressure and steadiness, the *chin* may be made to rest on the left hand. For the position of the worker see Plate IV. and V.
Trying-plane.	The manner of execution is shewn in No. 8 and in Fig. 91. The broader the surface, the more difficult is the exercise. To test whether the surface is really level, the plane is laid across it; or, better still, winding laths are laid, one on each end of the piece of work. If the upper edges lie in the same plane the surface is true.
File.	When the plane surface of an end piece is to be dressed up, the piece of wood should be secured in the carpenter's bench. If, on the other hand, the surface is convex, the work should merely be supported on it. In the former case the handle of the file is firmly grasped in one hand, whilst the other rests upon the back of the blade near the point. The tool is then passed steadily and evenly over the surface, pressure being exerted only when the file is going from the worker. If the work merely rests on the bench, the file can of course only be worked with one hand. When rounded surfaces are filed, the tool is worked in the direction of the fibres, or when this is impossible, obliquely to them.
Centre-bit.	For method of execution see No. 11. Care must be taken that the bit cuts evenly.

Number of Exercise.	Name of Exercise.	Purpose of Exercise.
15	*Convex sawing.*	To saw out a shape following a curved line.
16	*Concave cut.*	To cut out a concave shape. 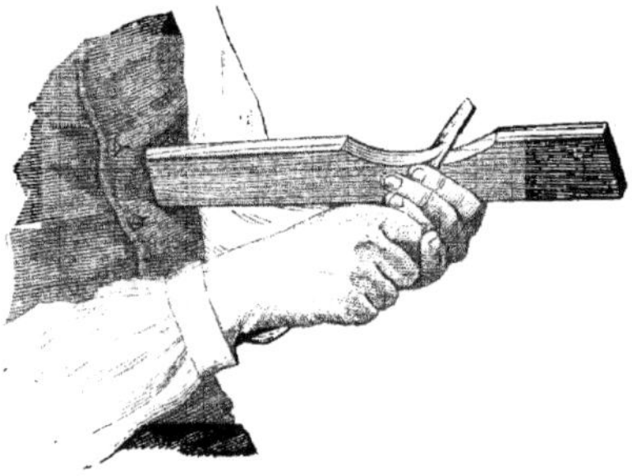 Fig. 94.
17	*Bevelling.*	To plane a bevelled edge.
18	*Convex modelling with the plane.*	To plane a convex surface.
19	*Sawing with tenon-saw.*	To saw carefully when no other saw can as advantageously be used.
20	*Wave-sawing.*	To saw out after a curved line. 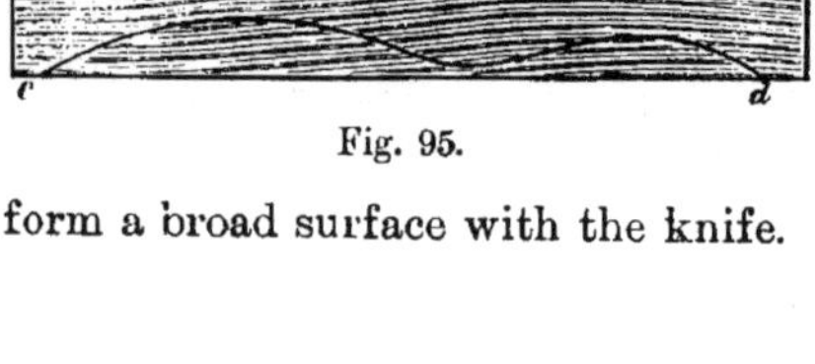 Fig. 95.
21	*Plane surface cut.*	To form a broad surface with the knife.

Tools required.	Directions for Work.
Turn-saw.	The saw is worked in the direction of a curve previously drawn (see Fig. 90, *a—b*).
Knife.	The knife is worked both from the worker and towards him, while the arm is pressed gently against the side, to steady the hand which holds the work. See farther No. 1.
Trying-plane.	The plane is made to produce a surface at an oblique angle to two others, in the same direction as the fibres. See farther No. 8.
Smoothing-plane.	The work is fastened between the bench-pegs. See No. 8 for method of execution.
Tenon-saw.	As the tenon saw has smaller teeth than the other saws used in slöjd carpentry, it is very suitable in cases where there is danger of splitting the wood. The tool should be worked with a light hand, and all pressure avoided.
Turn-saw.	For the method of execution see No. 7, bearing in mind that the frame of the saw must be inclined to the one side or to the other, according to the curves of the line (*c—d*, Fig. 95).
Knife.	Greater strength is required for this than for the preceding cuts, as almost the entire length of the blade is used.

Number of Exercise.	Name of Exercise.	Purpose of Exercise.
22	*Scraping.*	To finish up surfaces. 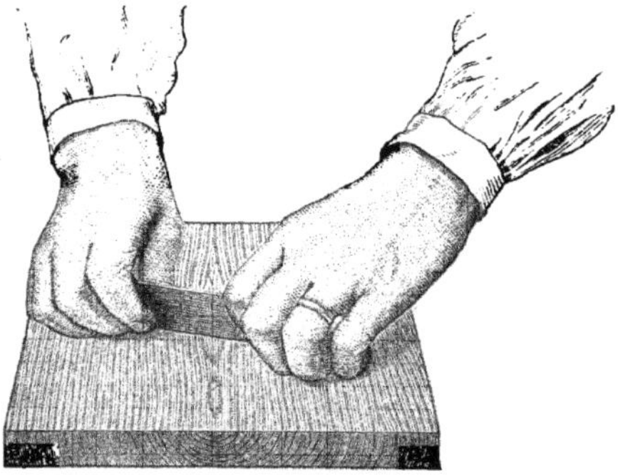 Fig. 96.
23	*Stop-planing (obstacle-planing).*	To plane a piece of wood which presents obstacles to the advance of the plane.  Fig. 97.
24	*Perpendicular chiselling, or paring.*	To cut down and smooth a surface.
25	*Oblique chiselling, or paring.*	To pare off a piece of wood obliquely to the fibres, but in the direction in which they run.

Tools required.	Directions for Work.
Scraper.	The tool is worked, as far as possible, in the direction of the fibres; in every other case obliquely to them. When the scraper is efficiently used, other means of finishing need only be sparingly employed.
Smoothing-plane.	The method of execution is, as nearly as possible, that described under No. 33, with this exception, that the tool is passed somewhat obliquely over the surface, in order to smooth it as near the obstacle as possible. [There is an English plane specially adapted for this sort of work called a *Stop Champher Plane.*—Trs.]
Firmer chisel.	The tool is grasped firmly by the handle in one hand, and worked perpendicularly, the upper arm being pressed firmly against the side to give the necessary support. The other hand holds the work on a cutting-board on the bench. (See Plate VI. for position of worker.)
Firmer chisel.	The piece of work must either be held firmly on the bench with one hand, or, when it seems necessary, fastened between the bench-pegs. The tool is firmly grasped by the other hand, and its face pressed against the wood (see Fig. 89, *c*—*d*). Oblique chiselling is always done in the direction in which the fibres run.

Number of Exercise.	Name of Exercise.	Purpose of Exercise.
26	*Gouging with the gouge and the spoon-iron.*	To produce depressions of various degrees of depth in a piece of wood.
27	*Concave chiselling.*	To produce a concave surface.
28	*Chopping.*	To split up and dress off rough and uneven surfaces.
29	*Smoothing or dressing up with the spokeshave.*	To dress up rounded surfaces. 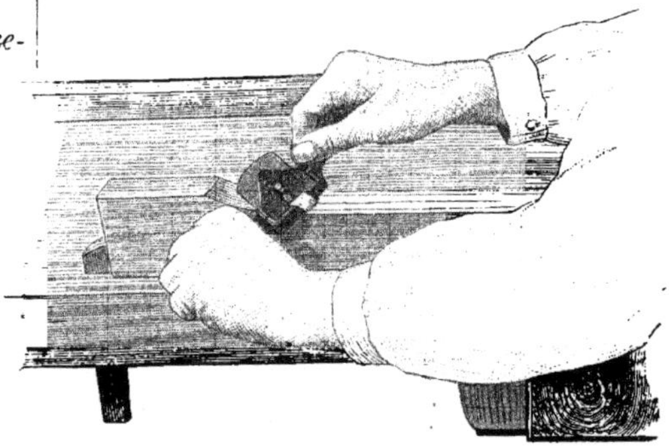 Fig. 98.
30	*Modelling with the spokeshave.*	To model rounded surfaces.
31	*Oblique sawing.*	To saw off a piece of wood obliquely to the fibres.

Tools required.	Directions for Work.
Gouge and spoon-iron.	The coarser preliminary work is done with the gouge, and the necessary pressure is given by mallet blows on the handle. The spoon-iron is worked with both hands, and the pressure thus given, being lateral, serves to remove the inequalities left by the gouge.
Firmer chisel.	The handle of the tool is firmly grasped in one hand, and the other hand rests upon the face of the blade to direct its course, which must always be in the direction in which the fibres run. The article should be made fast in the back bench-vice. (See Fig. 95, *a—b*.)
Axe.	One hand supports the piece of wood on the chopping-block; the other hand wields the axe. Should the wood be "contrary" it must be turned the way of the grain, or "humoured." For the position of the worker, see Plate VII.
Spokeshave.	The work is fixed in the bench-vice. The tool is firmly grasped in both hands, with heavy *forward* pressure from the thumbs, and *downward* pressure from the fingers. When necessary, the forward direction of working may be reversed. For the position of the worker see Plate VIII
Spokeshave.	For method of execution see No. 29.
Broad webbed bow-saw.	For method of execution see No. 5, and Fig. 89, *c—d*.

Number of Exercise.	Name of Exercise.	Purpose of Exercise.
32	*Oblique planing.*	To smooth surfaces in an oblique direction, over the fibres.
33	*Dressing with the smoothing-plane; or smoothing up.*	To produce a smooth and finished surface.

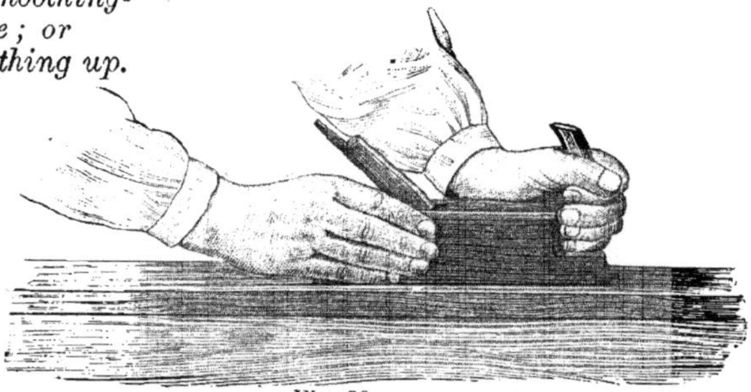

Fig. 99.

Number of Exercise.	Name of Exercise.	Purpose of Exercise.
34	*End squaring.*	To smooth up the surfaces of end pieces across the fibres.
35	*Halving with knife.*	To fasten two pieces of wood together as shown in Fig. 100.

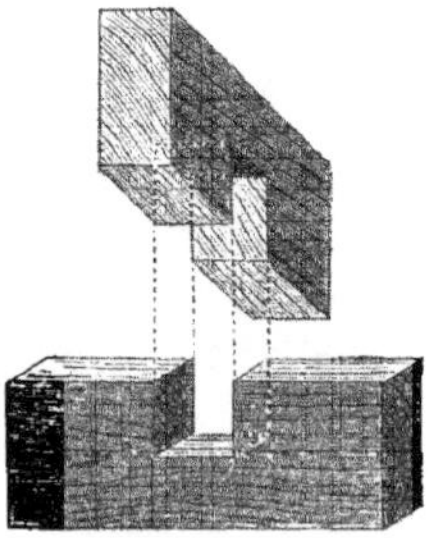

Fig. 100.

Number of Exercise.	Name of Exercise.	Purpose of Exercise.
36	*Work in hard wood.*	To manipulate very close-grained hard wood.

Tools required.	Directions for Work.
Smoothing-plane.	For method of execution see No. 33.
Smoothing-plane.	For directions for fastening the work into the bench see No. 8. The plane is firmly grasped in front and at the back (see Fig. 99), and worked briskly over the surface of the work. To produce a fine surface, the iron must be very sharp and lie as nearly as possible in the same plane as the sole, while the cover must lie close to the edge to prevent the fibres from splitting, whatever direction the plane may take. (See p. 100.)
Smoothing-plane.	The piece of wood is fastened vertically into the bench. To avoid splitting at the corners, the work should proceed from corners to centre.
Knife.	1. The work is set out with square, compass, and marking gauge. 2. It is cut out with the knife. 3. The parts are fitted together. This exercise requires great care and accuracy.
	As a general rule greater strength is required for wood of this kind than for softer wood.

Number of Exercise.	Name of Exercise.	Purpose of Exercise.
37 (*a*)	*Fitting in pegs.*	To joint two pieces of wood together by means of a dowel or pin which fits accurately into a hole bored with centre-bit or auger-bit.

Fig. 101.

Number of Exercise.	Name of Exercise.	Purpose of Exercise.
37(*b*)	*Plugging.*	To fill up a hole by means of a round plug.
38	*Bevelled edge-planing.*	To produce a plane surface at oblique angles to two other plane surfaces.

Fig. 102.

Tools required.	Directions for Work.
Square; centre-bit; knife.	1. The hole is drilled with the centre-bit. 2. After finding the centre of the peg, a circle is described with the bit to be used in order to get the size of the peg. 3. The dowel or pin is set out with the square and made cylindrical with the knife, so that it may fit closely and steadily into the hole.
Centre-bit; knife; firmer chisel.	The centre-bit used for the hole is used for marking out the size of the plug, and the fitting is done with the knife and chisel.
Trying-plane.	The exercise is performed according to the directions given in No. 17, with this exception, that the work is laid flat on the bench, and the plane is held obliquely at the angle required.

Number of Exercise.	Name of Exercise.	Purpose of Exercise.
39	*Glueing.*	To fix pieces of wood together.
40	*Boring with bradawl.*	To produce small holes in a piece of wood.
41	*Sinking and fixing metal plates, and other metal fittings.*	To fix simple metal plates, etc. on a piece of work.
42	*Nailing.*	To fasten pieces of wood together with nails.
43	*Punching.*	To sink a nail-head below the surface of the wood.
44	*Bevelling with the draw-knife.*	To produce a broad bevel in the same direction as the fibres run.

Fig. 103.

45	*Perpendicular gouging.*	To produce a concave excavation, perpendicular to a plane surface.

Tools required.	Purpose of Exercise.
	See Chap. IV. pp. 121, 122.
Bradawl.	The tool is worked with one hand, and turned steadily backwards and forwards under even pressure.
Screw-driver; firmer chisel; bradawl.	1. If the plate, etc. is to be sunk, the firmer chisel is used. 2. The plate is screwed on with the screw driver care being taken that the screw passes right down into the wood.
Hammer.	See jointing, Chapter IV., pp. 123, 124.
Punch; hammer.	The punch is held steadily on the head of the nail and struck sharply with the hammer; otherwise it may slip aside and make disfiguring holes in the work.
Draw-knife.	The article is screwed into the bench, and the tool, held firmly by both handles, is worked steadily over the wood. The exercise is rendered considerably easier if the draw-knife is held obliquely, *i.e.*, if the one end is somewhat in advance of the other. If contrary wood is encountered, the work should, if possible, be turned to allow the knife to cut in the direction of the fibres. [The face of the tool should be held towards the worker, with the bevelled edge on the work.—Trs.]
Gouge with edge on the convex side.	For method of execution see No. 24.

Number of Exercise.	Name of Exercise.	Purpose of Exercise.
46	*Plain jointing.*	To plane pieces of wood intended to be jointed by glueing.
47	*Dove-tail clamping.*	To insert a clamp in a broad piece of wood, to prevent warping.  Fig. 104.
48	*Oblique gouging.*	To pare a piece of wood in the direction of the fibres, but obliquely to them.
49	*Champhering.*	To pare a piece of wood at an obtuse or at an acute angle to its surface.

Tools required.	Directions for Work.
Trying-plane.	The greatest care is required in plain jointing in order that the surfaces which are to be united may fit accurately. The directions given in No. 8 must be followed. The trying-plane should be very finely set for this exercise, in order that it may remove very thin shavings. The angle which the edge makes with the side of the work must be frequently tested with the square; the straightness of the edge must also be tested by the eye. The second piece of wood is treated in the same way, and when it is ready, the edge of the first piece is placed upon it for trial. If the joint is accurate, the two surfaces will touch at all points, and when placed against the light will not allow a single ray to pass through. If any light shines through, the parts which are too high must be carefully planed down with long, steady strokes. If two or more planks are to be jointed for broad work, the surfaces must lie in the same plane, and this must be tested by placing against them the straight-edge or the blade of the square.
Compass; square; marking-point; bevel; marking-gauge; knife; tenon-saw; groove-saw; firmer chisel; old woman's tooth-plane; jack-plane; trying-plane.	1. The groove for the clamp is set out with the compass, square, marking-point, bevel, and marking-gauge; and a start for the saw is made with the knife. 2. It is cut out with the tenon-saw or groove-saw, firmer chisel, and old woman's tooth-plane. 3. The clamp is made ready with the jack-plane and the trying-plane, care being taken that it fits accurately all round.
Gouge.	For method of execution see No. 25, with this difference, that the gouge is used instead of the firmer chisel.
Firmer chisel.	For method of execution see No. 25.

Number of Exercise.	Name of Exercise.	Purpose of Exercise.
50	*Circular sawing.*	To saw out a circular shape.
51	*Screwing together, or fixing with screws.*	To fasten two pieces of work together by means of screws.
52	*Modelling with the draw-knife.*	To produce a rounded surface of large extent.
53	*Planing across the grain.*	To plane up a broad surface across the grain.
54	*Wedge planing with smoothing plane.*	To plane an article not only in the direction of the fibres, but obliquely to them (or to form an oblique object).
55	*Planing with round-plane.*	To dress up broad concave surfaces.
56	*Fixing with wooden pegs, for planing thin wood.*	To fix down, by means of wooden pegs, a thin piece of wood on the surface of a larger piece, in order to plane the former.

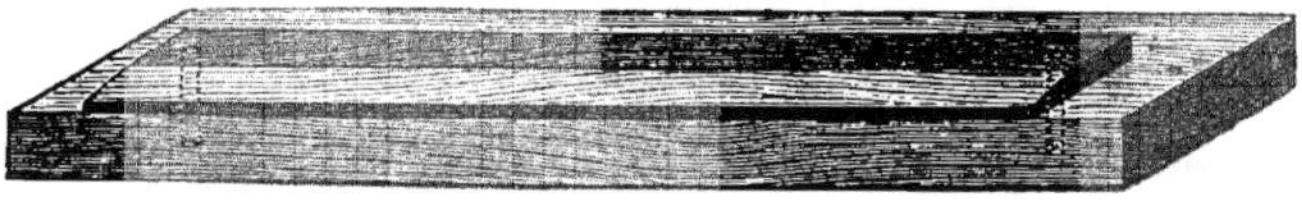

Fig. 105.

Number of Exercise.	Name of Exercise.	Purpose of Exercise.
57	*Single dove-tailing at right angles.*	To dove-tail two pieces of wood together by means of one dove-tail pin.

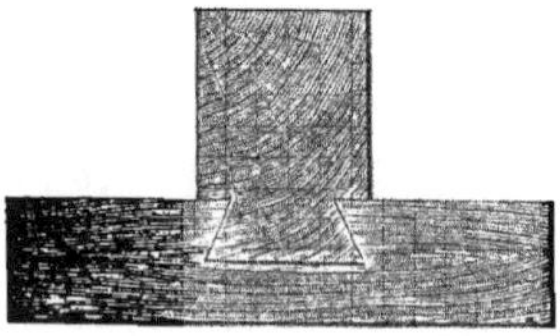

Fig. 106.

Tools required.	Directions for Work.
Turn-saw.	The piece of wood is fastened into the back bench-vice, and the saw is used according to the directions given in No. 15.
Screwdriver; pin-bit.	For method of execution see No. 41, and also jointing, Chap. IV., p. 124.
Draw-knife.	For method of execution see No. 44.
Trying-plane.	For method of execution see No. 12.
Smoothing-plane.	For method of execution see No. 33.
Round-plane.	For method of execution see No. 33. To produce a good result the plane must be worked very smoothly and steadily.
Pin-bit; knife; hammer.	Care must be taken that the under piece is level. The wood to be planed is placed on it, and holes are drilled with the pin-bit, close to each end of the upper piece, *through to the under piece.* Suitable pins are then driven into these holes, and a stable foundation is thus provided for the work of the plane.
Compass; square; marking-point; bevel; cutting-gauge; knife; tenon-saw; groove-saw; firmer chisel; old woman's tooth-plane.	1. The groove is set out with compass, square, marking point, bevel and cutting-gauge, and a start for the saw is made with the knife. 2. It is cut out with the tenon-saw or groove-saw, firmer chisel, and old woman's tooth-plane. 3. The clamp is set out with the cutting-gauge, and cut out with the knife. Care must be taken that the dove-tail fits accurately.

Number of Exercise.	Name of Exercise.	Purpose of Exercise.
58	*Common dove-tailing.*	To "corner-joint" by dove-tailing, *i.e.*, to insert bevelled pins into tightly fitting sockets.

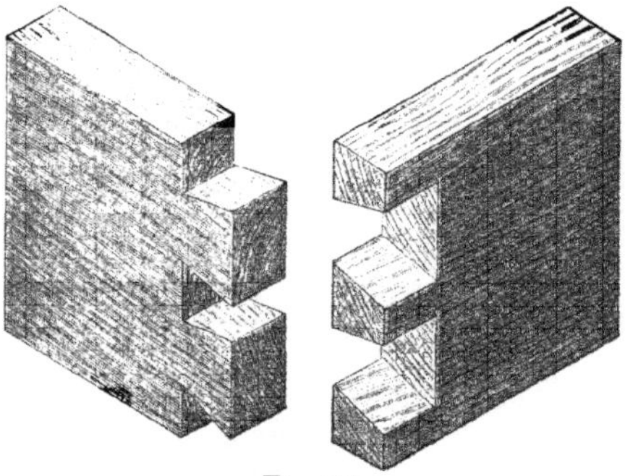

Fig. 107.

Number of Exercise.	Name of Exercise.	Purpose of Exercise.
59	*Square shooting, or planing with shooting-board.*	To plane a narrow piece of wood across the grain by means of the shooting-board.

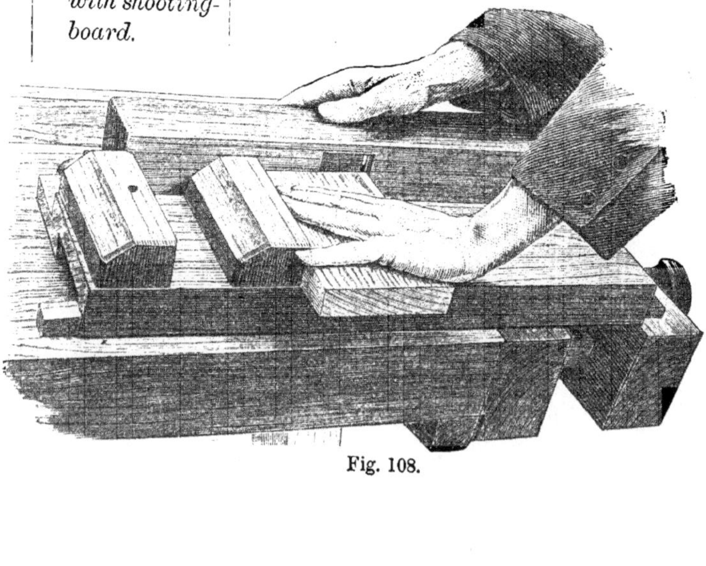

Fig. 108.

Tools required.	Directions for Work.
Cutting-gauge; compass; bevel; square; dove-tail saw; marking-point; firmer chisel.	1. The thickness of the wood to be dove-tailed is marked with the cutting-gauge across the ends of the pieces of wood on both sides. 2. The required bevel of the pins is indicated with compass, bevel, and square. 3. The pins are cut out with dove-tail saw and firmer chisel. 4. The pin end is held steadily on the other piece of wood at right angles to it, and the pins are marked out with the marking point. Then these marks are squared across the end of the wood. 5. They are cut out with the dove-tail saw and the firmer chisel. 6. The parts are carefully fitted together.
Trying-plane; shooting-board.	For the proper setting of the plane see No. 8. Great care is necessary when the shooting-board is in use, because the worker may easily hurt himself. See p. 67.

Number of Exercise.	Name of Exercise.	Purpose of Exercise.
60	*Hollowing out; or scooping out with gouge.*	To produce narrow concave depressions; or to hollow out with the gouge.
61	*Axle fitting.* [*This exercise only applies to one Swedish model, i.e., the shuttle.*—TRS.]	To fit an axle into a hole.
62	*Housing, or square grooving.*	To divide boxes, etc. into two or more rectangular portions by means of pieces of wood. 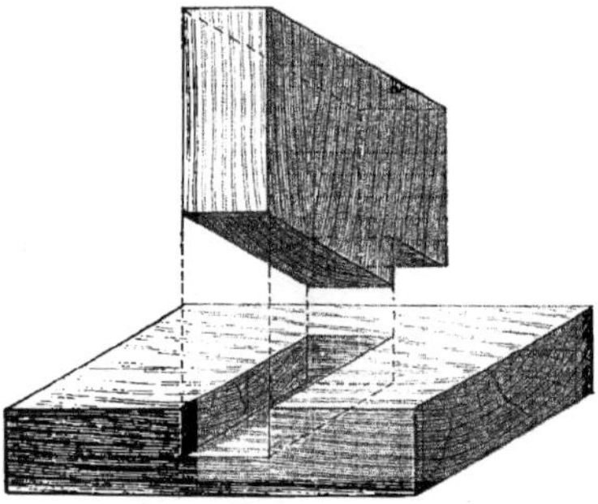 Fig. 109.
63	*Long oblique planing.*	To plane a long bevelled edge.
64	*Setting out.*	To set out divisions in the work.

Tools required.	Directions for Work.
Gouge.	For method of execution see No. 26.
Compass; bradawl; firmer chisel.	1. The axle is set out with the compass. 2. The hole and the slot are made with the bradawl and chisel.
Compass; square; marking-gauge; saw; firmer chisel; smoothing-plane.	1. The groove is set out by means of the compass, square, and marking-gauge. 2. It is cut out with the saw and the firmer chisel. 3. The tenon, *i.e.*, the piece which is set into the groove, is made to fit by means of the smoothing-plane.
Trying-plane.	For method of execution see No. 8.
Compass; marking-point; square.	1. The length is divided with the compass, first into larger, and then into smaller parts. 2. The lines are drawn with the marking-point at right angles to the edge of the object. Great accuracy is required in marking off the divisions

Number of Exercise.	Name of Exercise.	Purpose of Exercise.
65	*Panel-grooving.*	To produce rectangular depressions in an object, into which a flat piece of wood is to be slotted. 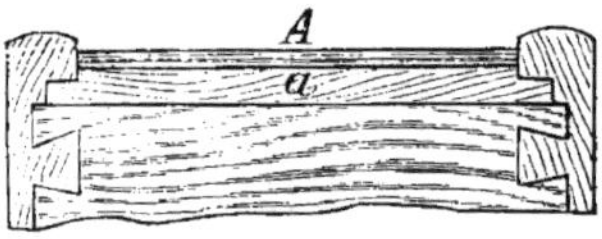 Fig. 110.
66	*Glueing with aid of hand-screw.*	To glue together with the aid of the hand-screw.
67	*Sawing with compass (or keyhole) saw.*	To saw out a hole in a piece of wood.
68	*Oblique edge-grooving.*	To join two pieces of wood together by means of a single dove-tail, at an obtuse angle. 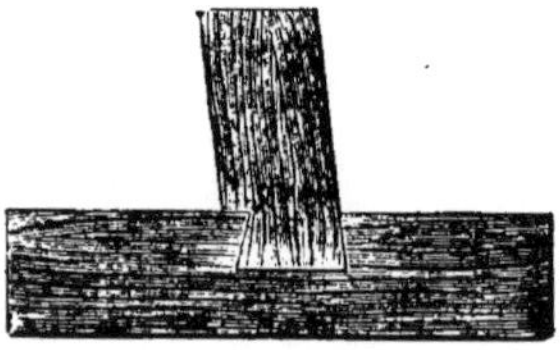 Fig. 111.

Tools required.	Directions for Work.
Cutting-gauge; knife; firmer chisel; plough.	1. The groove is set out with the cutting-gauge. 2. It is cut out with the knife and the firmer chisel. In the case of many objects the plough may be employed with advantage to cut out the groove.
Handscrew.	Before the glue is applied to the joint, the parts must fit accurately; otherwise the pressure of the handscrew will be of little service. A piece of wood should be laid between the work and the screw to prevent injury to the surface of the article, and also to distribute the pressure more equally.
Centre-bit; compass saw.	Two holes are drilled in the piece of work with the centre-bit. The article is then fastened vertically into the bench and the saw is worked from one hole to the other, following lines previously set out. (In the case of small articles, use may be made of a turn-saw, the blade of which is detachable at one end.)
Compass: square; marking point; bevel; cutting-gauge; knife; tenon-saw or groove-saw; firmer chisel; old woman's tooth-plane; smoothing-plane.	1. The groove is set out with the compass, square, marking-point, bevel, and cutting-gauge; and a start for the saw is made with the knife. 2. It is cut out with the tenon-saw or groove-saw, firmer chisel, and old woman's tooth-plane. 3. The required form of the end of the dove-tail is set out with the cutting-gauge and the bevel, and it is bevelled with the smoothing-plane. 4. The clamp is cut out with the knife. 5. The parts are fitted together.

Number of Exercise.	Name of Exercise.	Purpose of Exercise.
69	*Slotting.*	To join two pieces of wood, of which one is thinner than the other, in such a manner that the former slots into the latter at a right angle. 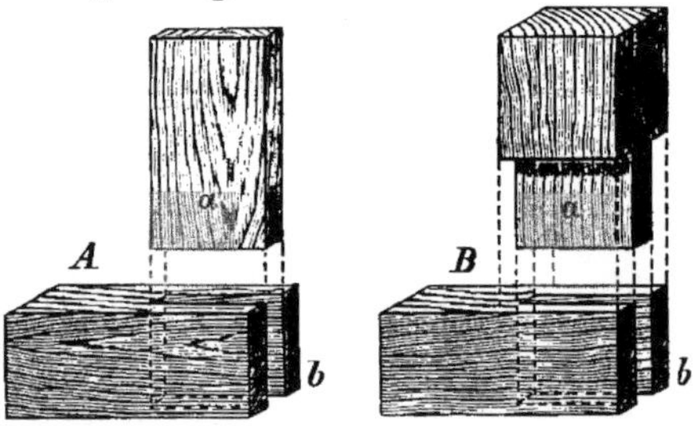 Fig. 112.
70	*Dove-tailing in thick wood.*	To make a rectangular corner-joint by dove-tailing two pieces of thick wood.
71	*Mitreing.*	To make an end-joint with two pieces of wood at an angle of 45°.  Fig. 113.
72	*Common mortise and tenon.*	To join by means of a mortise and tenon.

Tools required.	Directions for Work.
Square; marking gauge; tenon-saw (or dove-tail saw) firmer chisel; mallet.	1. The tenon (*A a*) and (*B a*), and the slot (*B b*), are set out with the square and the marking gauge. 2. The slot (*B b*) is cut down with the tenon (or dove-tail) saw, and cut out with a coarse firmer chisel, (or mortise-chisel) by aid of the mallet. 3. The tenon (*B a*) is made with the tenon (or dove-tail) saw and the firmer chisel. It is called a shoulder tenon. The tenon (*A a*) is simply fitted into the slot with the smoothing plane. It is called an unshouldered tenon. 4. The parts are fitted together, and if necessary the firmer chisel is used.
Cutting-gauge; compass; bevel; square; dove-tail saw; marking-point; firmer chisel.	For method of execution see No. 58; but note that still greater accuracy is required, because exercises with the saw and the firmer chisel are always more difficult when the thickness of the wood either falls under a certain limit, or exceeds it (Fig. 107).
Square; compass; firmer chisel; smoothing-plane.	For method of execution see No. 25. The completion of the joint depends on the nature of the object in which the exercise occurs. It may require mortising, glueing, nailing, screwing together with wood screws, etc. When the object is large, the smoothing-plane is used in mitreing. [The English method of making this mitre is by means of a mitre-box and shooting-board, in which case the saw and trying-plane are used.—Trs.]
Square; mortise gauge; mortise chisel; mallet; tenon-saw.	1. The mortise is set out by means of the square and the mortise gauge. 2. It is cut out with the mortise-chisel with the aid of the mallet. 3. The tenon is cut out with tenon-saw and firmer chisel. 4. The parts are fitted together, the firmer chisel being employed when necessary.

Number of Exercise.	Name of Exercise.	Purpose of Exercise.
73	*Half-lapping.*	To joint together two pieces of wood by half-lapping the broad sides together, *i.e.*, by cutting half the depth of the wood away from each. 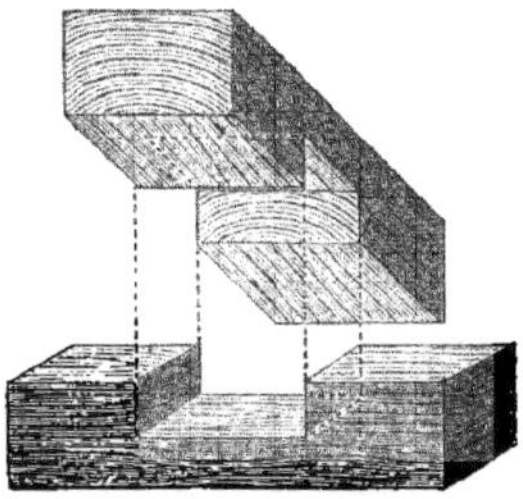 Fig. 114.
74	*Rebating.*	To make a rebate.  Fig. 115.
75	*Graving with V-tool or parting-tool (fluting).*	To hollow out depressions or edges.

Tools required.	Directions for Work.
Square; marking point; marking-gauge; tenon-saw; firmer chisel.	1. The half-lapping parts are set out with the square, marking-point and marking-gauge. 2. They are cut out with tenon-saw and firmer chisel. 3. The parts are fitted together with the aid of the chisel.
Marking-gauge; knife; firmer chisel.	1. The breadth and thickness of the rebate are set out with the marking-gauge. 2. It is cut out with the knife and the firmer chisel. [This holds good only of the small rebates which occur in Slöjd carpentry. The plough and the rebate plane are used for larger work.—Trs.]
Parting-tool.	The object is screwed tightly into the bench, and the parting-tool is wielded with a steady hand.

Number of Exercise.	Name of Exercise.	Purpose of Exercise.
76	*Half-lap dove-tailing.*	To produce a rectangular end joint by dove-tailing together two pieces of wood, so that the dove-tailing does not show on one side. To do this, one-third of the wood is not cut through on the side where the pins are. The socket piece is cut right through and dove-tailed in the ordinary way. 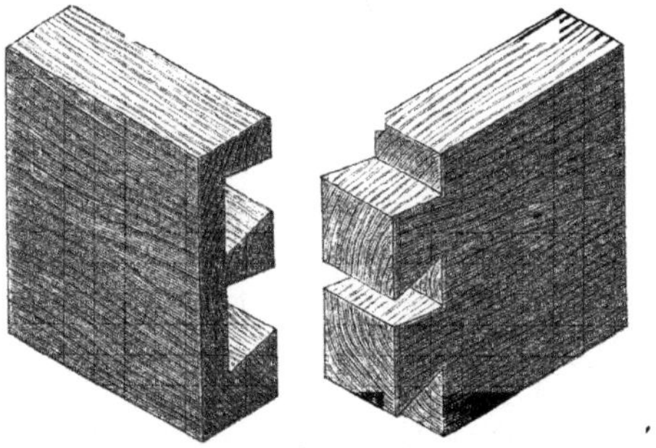 Fig. 116.
77	*Hinge-sinking, or fixing hinges.*	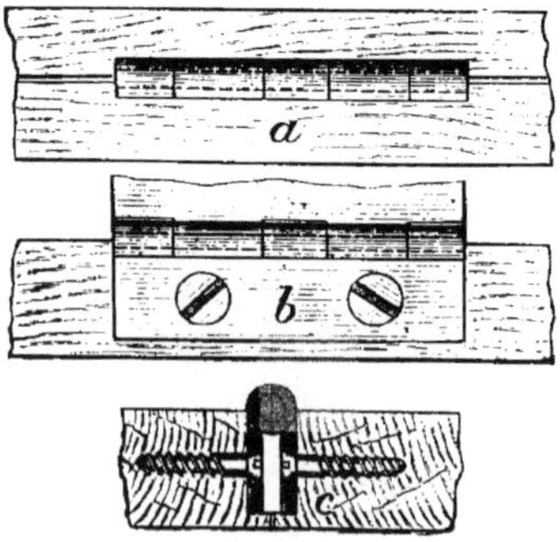 Fig. 117.
78	*Lock-fitting.*	

Tools required.	Directions for Work.
Cutting-gauge; compass; bevel; square; tenon-saw (dove-tail saw); marking point; firmer chisel.	For method of execution see No. 58, paying special attention to setting the piece on which the pins are in an oblique position in the back bench-vice. The pins are sawn out to the lines indicated by the marking point, which determine the thickness of the wood to be left. The spaces between are smoothed by perpendicular paring with the firmer chisel. (See No. 24.)
Square; firmer-chisel; bradawl; screwdriver.	1. The position of the hinge is decided on and set out. 2. The depth to which the hinge has to be sunk is taken and gauged. 3. This part is then cut out with the firmer chisel. 4. For screwing on, see No. 41.
Pin-bit; firmer chisel; knife; bradawl; screwdriver; compass-saw.	1. The position of the lock is decided on. 2. The place is cut out with the pin-bit and the firmer chisel to a depth which permits the metal plate to lie on the same plane as the wood. 3. The hole for the key is cut out with the centre-bit, knife and chisel. (In larger work with the compass-saw). 4. For screwing on, see No. 41.

Number of Exercise.	Name of Exercise.	Purpose of Exercise.
79	*Oblique dove-tailing.*	To make a rectangular end-joint with oblique pieces of wood. 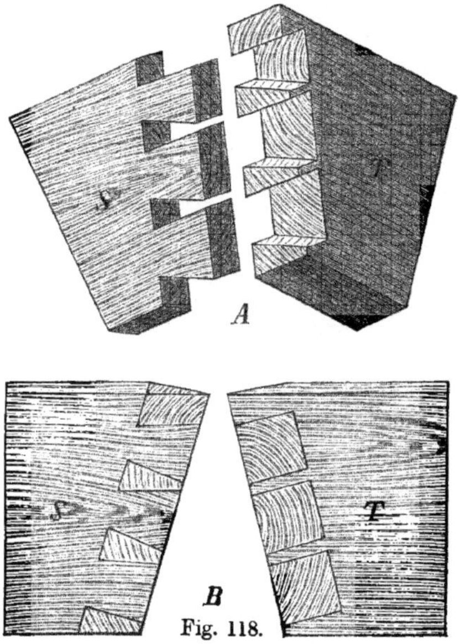 Fig. 118.
80	*Oblique slotting*	To make an oblique angled joint with a pin and a slot.

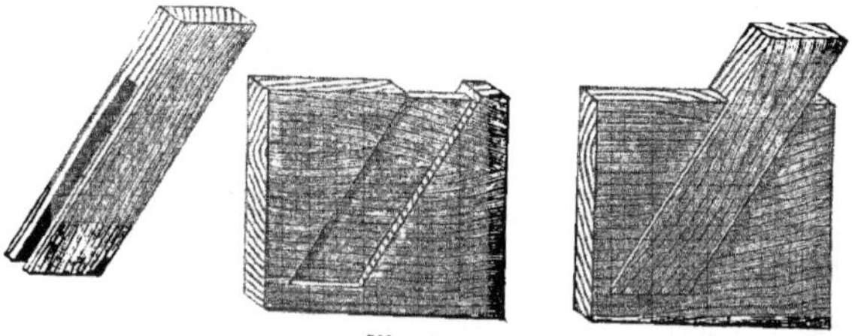

Fig. 119.

Tools required.	Directions for Work.
Bevel; dove-tail saw; compass; square; tenon saw; firmer chisel; smoothing-plane.	1. Set out the angle at the ends with the bevel and saw off. 2. Bevel off the edges to correspond with the angle at the ends of the adjacent sides. 3. To get the angle in the *horizontal* plane at the ends, use the square in the following way: Place the face against the side of the wood, and let the blade rest flat on the plane of the bevelled edge. Then draw the line and plane off. 4. The required thickness of each piece is set out with the cutting gauge. 5. The pins are set out at right angles to the oblique end with the compass, bevel, and square. 6. They are made with the dove-tail saw and the firmer chisel. [Another method of working this joint is by means of a prepared shooting board, by which the two angles at the ends can be obtained at once. It may also be mentioned, that in the English method of oblique dove-tailing, the dove-tail pins run in the same direction as the grain, or obliquely to it, and are consequently stronger. There are theoretical reasons why this method is not followed at Nääs.—Trs.]
Compass; square; bevel; marking-gauge; knife; firmer chisel; dove-tail saw.	1. The slot is set out with compass, square, and bevel. 2. The depth of the groove is set out with the marking-gauge, and cut out with the knife and firmer chisel. 3. The slot is sawn out with the dove-tail saw, and cut out with the firmer chisel. 4. The parts are fitted together with the aid of the firmer chisel.

Number of Exercise.	Name of Exercise.	Purpose of Exercise.
81	*Notched dove-tailing (half concealed edge grooving).*	To insert a dove-tail, the outer edge of which conceals the groove, into a piece of wood.

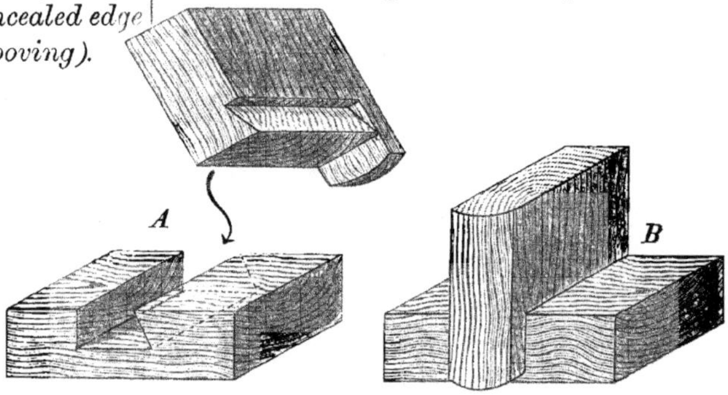

Fig. 120.

Number of Exercise.	Name of Exercise.	Purpose of Exercise.
82	*Concave modelling with plane (hollowing with plane).*	To produce a concave surface with the plane.
83	*Staving.*	To fix concave-shaped pieces of wood or staves to a curvilinear bottom piece, to make a barrel or bucket.

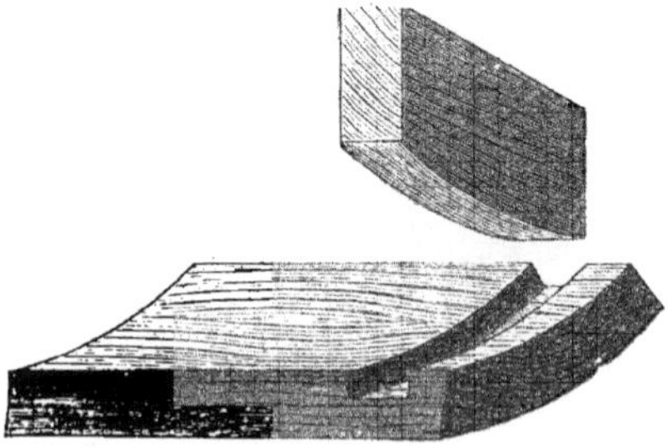

Fig. 121.

Tools required.	Directions for Work.
Compass; square; marking point; bevel; cutting gauge; knife; tenon-saw or groove-saw; firmer chisel; old woman's tooth plane; dove-tail saw.	1. The groove is set out with compass, square, marking-point, bevel, and cutting gauge; cut out with knife, tenon-saw or groove-saw, firmer chisel, and old woman's tooth-plane 2. The shape of the dove-tail is set out with square and cutting gauge, and cut out with dove-tail saw and knife. 3. The parts are fitted together with the aid of the knife.
Round plane.	1. The required curve is set out at both ends. 2. The shape is produced by means of the roughing plane and the "round" plane. (See Fig. 121.)
Compass; bow-saw; spoke-shave; bevel; marking-point; marking gauge; knife; firmer chisel; trying-plane; smoothing-plane; brad-awl.	1. The bottom is made in the shape required. 2. The edge of the bottom is bevelled to the angle required for the sides of the article. 3. The position and breadth of the groove are set out with the marking-point. 4. The necessary inclination of the sides of the staves is determined by the bevel. 5. The depth of the groove is set out with the marking-gauge and cut out with the knife and the firmer chisel. 6. The edges of the staves are planed and fitted together. 7. The staves are held together by means of wooden pins inserted into the edges from the inside.

Number of Exercise.	Name of Exercise.	Purpose of Exercise.
84	*Hooping.*	To fix iron hoops round a barrel or bucket, to hold the staves together. (The wooden hoops frequently seen are not suitable for slöjd work.)
85	*Concealed tenoning.*	To joint two pieces of wood together at right angles by means of a concealed or haunched tenon.

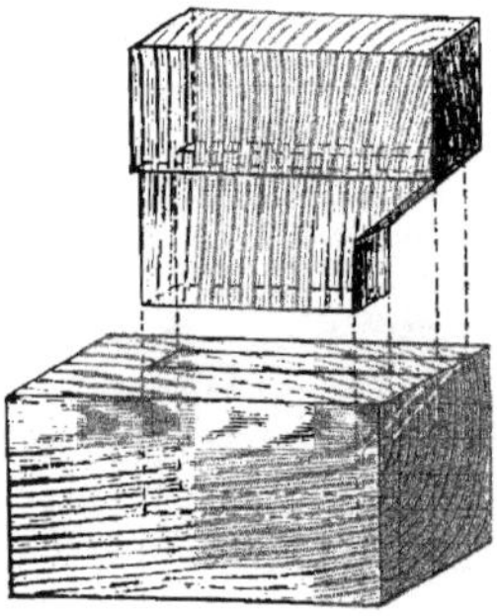

Fig. 122 a.

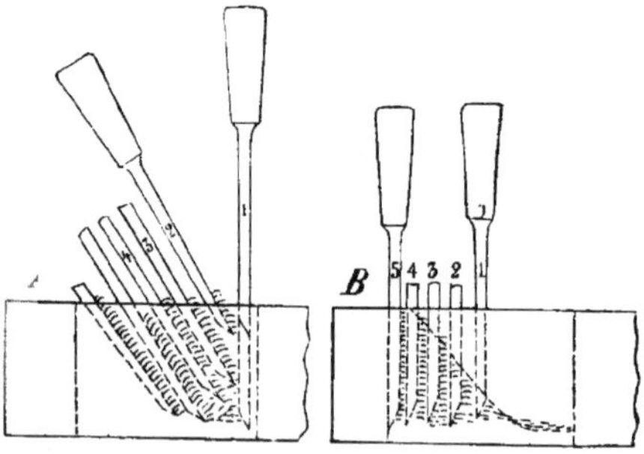

Fig. 122 b.

Tools required.	Directions for Work.
Cold chisel; punch; hammer; set hammer.	1. The length of the hoop is taken and cut off with the cold chisel. 2. A hole is made with the punch and hammer about half an inch from each end. 3. The hoop is rivetted by means of a rivet, the head of which is larger than the hole. 4. The head of the rivet is then made to rest on a block of metal, and the rivet itself is hammered until a head is formed on the other side. 5. The hoop is hammered from the inside of the article as it rests on the block, and thus made to fit. 6. The hoops are put on from the narrowest portion of the article, and driven home by blows from the set hammer.
Square; mortise gauge; firmer chisel; mortise chisel; mallet; bow-saw; tenon-saw.	1. The tenon and the mortise are set out at right angles. 2. The breadth of the mortise and the thickness of the tenon are set out with the mortise gauge. 3. The mortise is cut out with the mortise chisel. 4. The tenon is made with the bow-saw, firmer chisel, and tenon-saw. 5. The parts are fitted together with the help of the firmer chisel.

Number of Exercise.	Name of Exercise.	Purpose of Exercise.
86	*Blocking.*	To strengthen by means of blocks. N.B.—In the illustration the fibres of the block are accidentally shown running in the wrong direction. 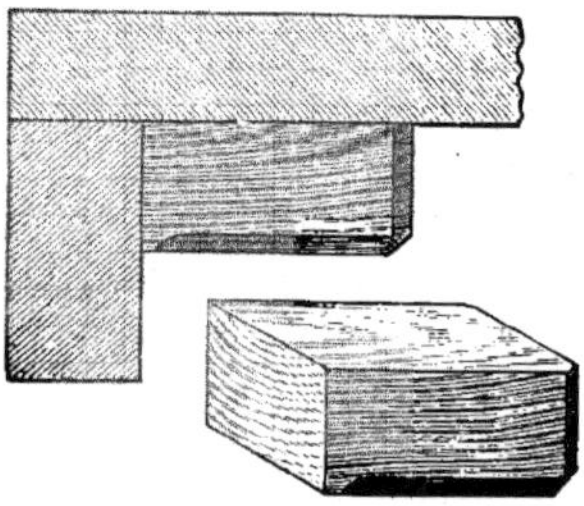 Fig. 123.
87	*Mortised blocking.*	To strengthen by means of mortised blocks. [Sometimes called "button blocks."—Trs.]  Fig. 124.
88	*Up and down sawing.*	To divide a long piece of wood into small pieces.

Tools required.	Directions for Work.
Dove-tail saw; firmer chisel; square.	1. The piece to be strengthened is held close to the other piece with the handscrew. 2. The blocks are made with the dove-tail saw and the chisel. 3. The blocks are warmed, glued, and put into their places. 4. Before the handscrew is taken away, the glue must be quite dry.
Square; marking gauge; firmer chisel; dove-tail saw.	1. The mortise in the rail is set out at right angles with the square and the marking-gauge, and cut out with the firmer chisel. 2. The tenon on the block is set out in accordance with the size of the mortise with the square and the marking gauge; cut out with the dove-tail saw, and fitted with the firmer chisel. 3. Previous to the blocking, the object to be fixed is held in position with the handscrew. 4. The blocks are warmed and glued on the two sides next the object, and put in their places.
Broad-webbed bow-saw.	The plank is placed on the bench and held in place by a handscrew. The blade of the saw is set almost at right angles to the plane of the frame, and the handle is grasped by one hand, while the other holds the upper end of the side arm, and the saw is worked vertically with long easy strokes, with the blade at right angles to the plane surface of the plank.

Plate I. Position: Convex Cut.

Plate II. Position: Long-sawing.

Plate III. Position: Edge-planing.

Plate IV. Position: Perpendicular boring with the brace.

Plate V. Position: Horizontal boring with the brace.

Plate VI. Position: Perpendicular chiselling.

Plate VII. Position: Chopping.

Plate VIII. Position: Smoothing, &c., with the spokeshave.

Plate IX. Plan of Slöjd-room in Katarina Elementary School, Stockholm.

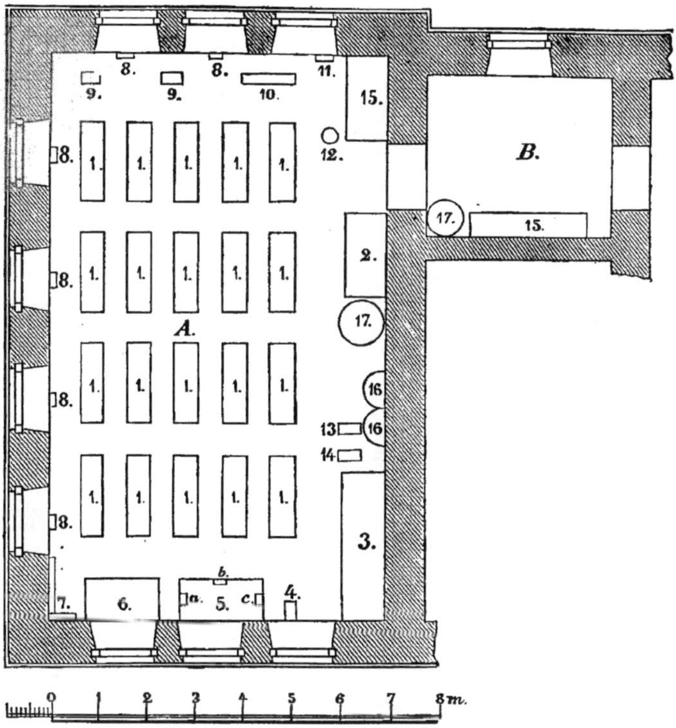

A. Slöjd room.
1. Benches.
2. Cupboard with two divisions: (1) for tools; (2) for models.
3. Cupboard with two divisions: (1) for unfinished work. (2) for finished articles.
4. Teacher's desk.
5. Cupboard:
   a. Iron vice.
   b. Iron saw-file vice.
   c. Anvil.
6. Lathe.
7. Racks for hand-screws and shooting-boards.
8. Vices for rough work.
9. Boring-stools.
10. Saw-bench.
11. Glue-pot.
12. Chopping-block.
13. Flat grindstone.
14. Revolving grindstone.
15. Wood-racks.
16. Wash-hand basins.
17. Stoves.
    3 racks for saws are introduced between the windows on the long wall.

B. Lobby.

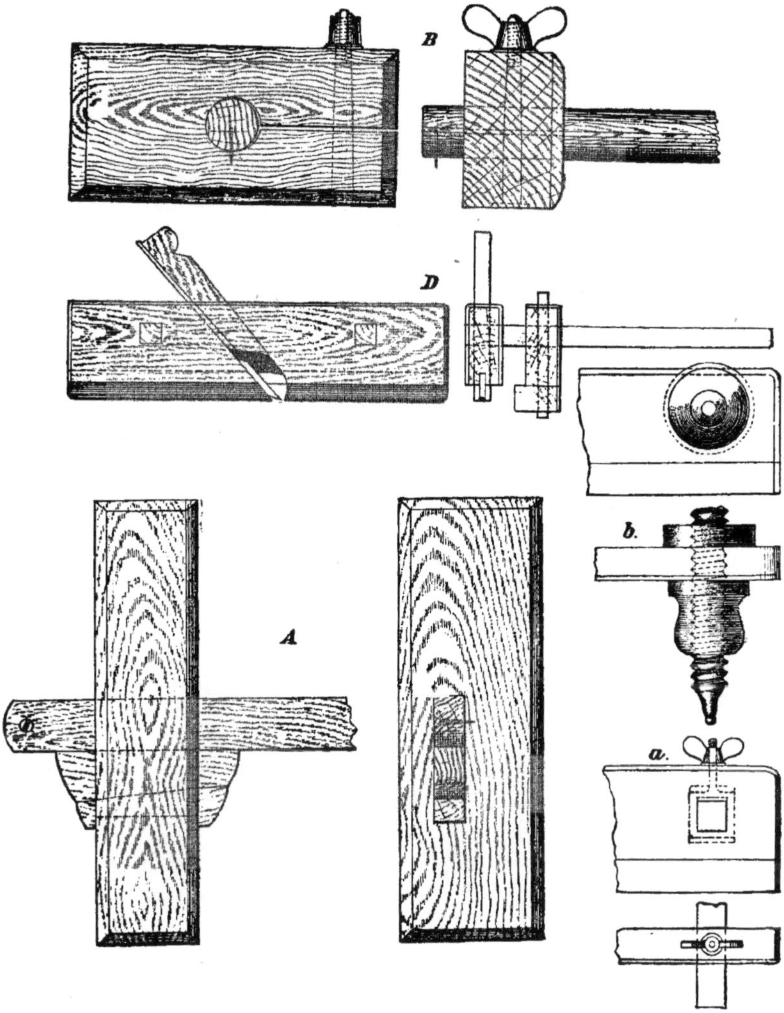

Plate X.

A. Marking gauge (Johansson's) with stock adjusted by wedges. ½.
B. Marking gauge (Lundmark's modified) with stock adjusted by thumb-screw. ½.
D. Plough-gauge, *a* and *b*, different methods of adjustment. ⅓.

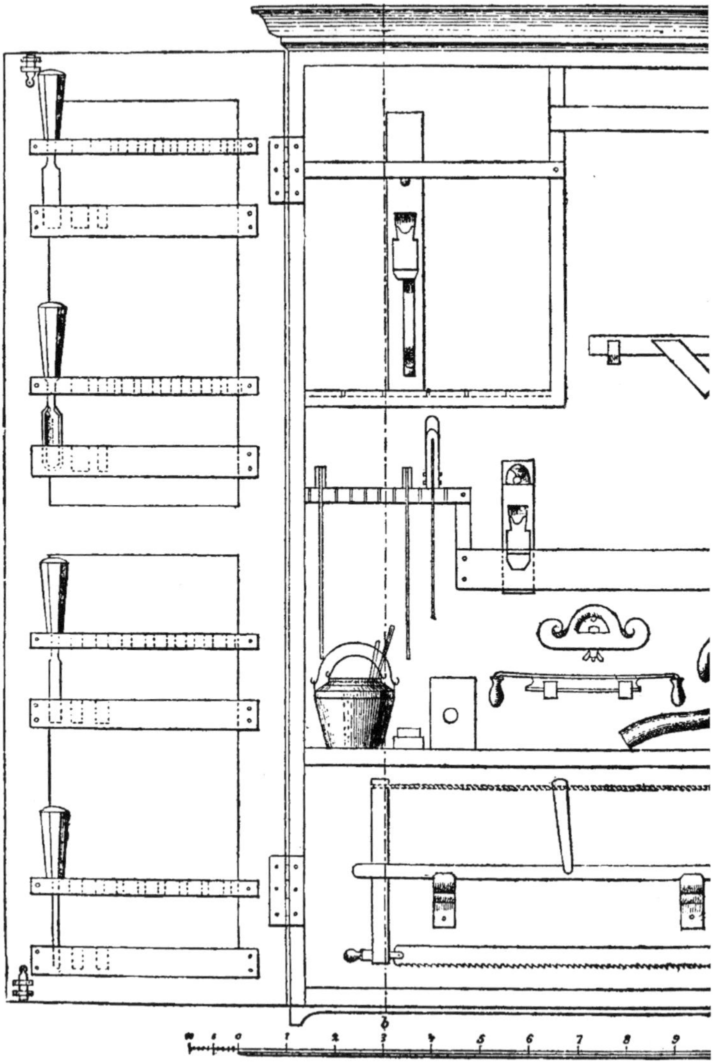

Plate XI. Tool Cup

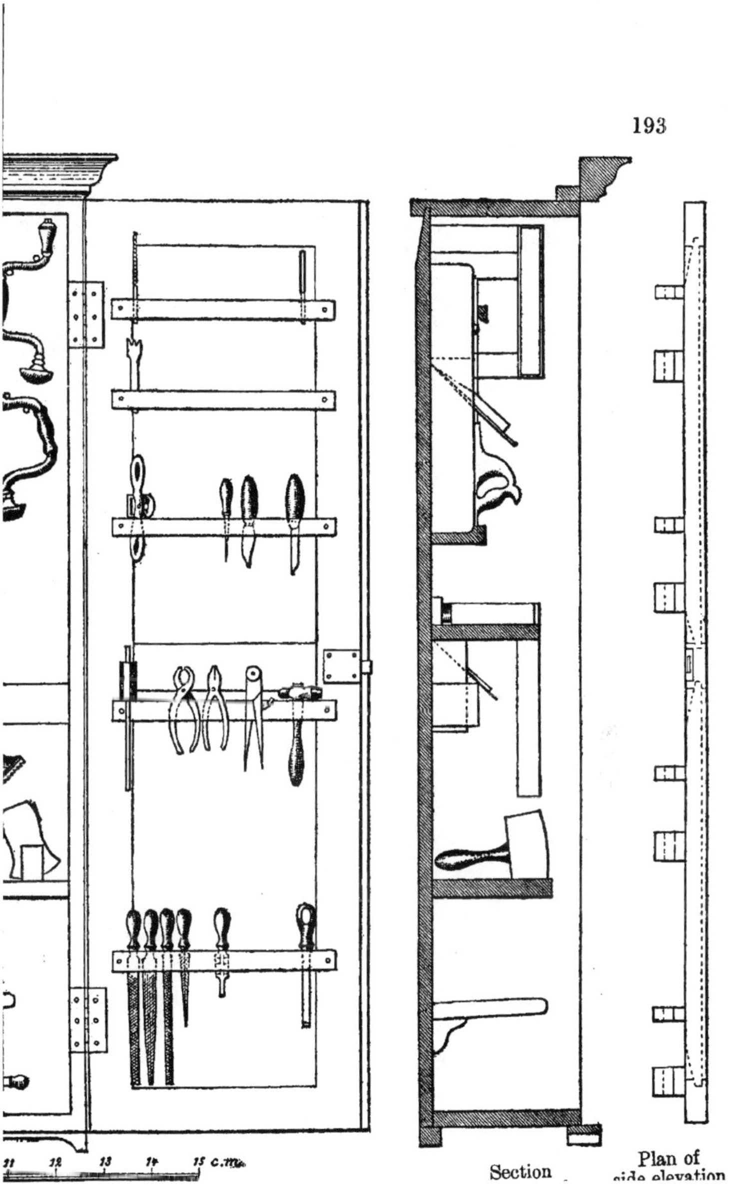
11 12 13 14 15 c.m.
Section
Plan of
side elevation

The Table given below represents a systematically arranged Series of Models (the High School Series). The numbers in the widest column indicate the Exercises required in each Model, in the order in which they occur.

	Name of Model.	Wood employed.	Exercises involved.	New Exercises.	Total number of Exercises performed.
1*a*	Small pointer	Birch	1. 2.	1. 2.	2
1*b*	Do.	Do.	1. 2. 3.	3.	3
2	Parcel pin	Do.	1. 2. 4.	4.	4
3	Round flower stick	Pine	5. 1. 2. 6.	5. 6.	6
4	Letter opener	Birch	5. 1. 2. 6.	0.	6
5	Rectangular flower stick	Pine	5. 7. 8. 9. 3. 6.	7. 8. 9.	9
6	Pencil holder	Do.	5. 1. 11. 6. 2. 13.	11. 13.	11
7	Key label	Do.	5. 7. 12. 8. 9. 10. 11. 6. 2. 13. 3.	10. 12.	13
8	Packthread winder	Birch	5. 7. 12. 8. 9. 14. 10. 15. 1. 6. 16. 13.	14. 15. 16.	16
9	Round ruler	Pine	5. 7. 8. 9. 10. 17. 18. 6. 2. 13.	17. 18.	18
10	Pen-rest	Birch	5. 28. 8. 9. 10. 19. 1. 2. 18. 13.	19. 28.	20
11	Paper knife	Do.	5. 7. 12. 8. 9. 10. 20. 16. 6. 21. 13. 22.	20. 21. 22.	23
12	Razor strop	Do.	5. 7. 12. 8. 9. 10. 23. 24. 14. 13. 22.	23. 24.	25
13	Small bowl	Do.	5. 7. 12. 8. 9. 10. 15. 24. 13. 26. 25. 6. 22.	25. 26.	27
14	Hammer handle	Do.	5. 7. 12. 8. 9. 20. 29. 4. 30. 2. 13. 22.	29. 30.	29
15	Pen tray	Do.	5. 7. 12. 8. 9. 10. 26. 22. 24. 13. 18. 33.	33.	30
16	Knife board	Pine or fir	5. 12. 8. 9. 10. 14. 15. 24. 34. 13. 33. 22.	34.	31
17	Flower-pot cross	Pine	5. 7. 8. 9. 10. 2. 1. 6. 13. 35.	35.	32

	Name of Model.	Wood employed.	Exercises involved.	New Exercises.	Total number of Exercises performed.
18	Metre measure	Birch	5. 7. 12. 8. 9. 10. 23. 29. 15. 16. 1. 6. 2. 4. 13. 22.	0.	32
19	Scoop	Do.	5. 28. 12. 9. 14. 7. 15. 24. 31. 32. 25. 26. 13. 18. 6. 16. 2. 22.	31. 32.	34
20	Clothes rack	Pine	5. 7. 12. 8. 9. 10. 14. 15. 24. 13. 38. 4. 2. 6. 37. 25. 33. 39. 41. 22.	37. 38. 39. 41.	38
21	Flower pot stand	Do.	5. 7. 12. 8. 9. 10. 34. 19. 24. 1. 2. 13. 42. 43. 33.	42. 43.	40
22	Flower-press roller and rests	Birch	5. 7. 12. 9. 10. 14. 17. 18. 24. 13. 22. 8. 15. 45. 32. 11.	45.	41
23	Footstool	Pine	5. 7. 12. 8. 9. 10. 34. 15. 45. 31. 25. 14. 13. 33. 42. 43.	0.	41
24	Book-carrier	Pine and birch	5. 12. 8. 9. 47. 39. 10. 34. 19. 24. 1. 15. 27. 2. 18. 13. 4. 22. 11. 51.	27. 47. 51.	44
25	Box	Pine	5. 7. 12. 8. 9. 10. 59. 42. 43. 33. 22.	59.	45
26	Ladle	Birch	5. 28. 12. 9. 10. 31. 15. 32. 25. 14. 24. 26. 20. 52. 30. 1. 6. 16. 13. 22.	52.	46
27	Flower-press	Pine	5. 28. 12. 46. 39. 9. 47. 15. 10. 29. 53. 54. 33. 22. 51.	46. 53. 54.	49
28	Cloak-suspender	Birch	5. 7. 12. 8. 9. 10. 15. 55. 33. 2. 6. 14. 11. 30. 13. 22. 37. 41.	55.	50
29	Flat ruler	Do.	5. 7. 12. 8. 9. 10. 14. 11. 56. 38. 2. 13. 22.	56.	51
30	Boot-jack	Birch and Pine	5. 7. 12. 8. 9. 31. 15. 10. 29. 27. 57. 39. 34. 32. 30. 16. 6. 2. 13. 22.	57.	52

	Name of Model.	Wood employed.	Exercises involved	New Exercises.	Total number of Exercises performed.
31	Lamp wall stand	Pine	5. 7. 12. 8. 9. 10. 59. 58. 15. 16. 1. 6. 2. 13. 14. 40. 39. 42. 43. 33. 22.	40. 58.	54
32	Shuttle	Birch	5. 7. 12. 8. 9. 15. 29. 24. 60. 10. 6. 1. 40. 61. 13. 22.	60. 61.	56
33	Knife-box	Pine	5. 7. 12. 8. 9. 10. 59. 58. 62. 39. 14. 20. 6. 16. 2. 1. 13. 46. 33. 18. 42. 43. 22.	62	57
34	American axe-handle	Oak	5. 20. 12. 10. 29. 44. 30. 18. 13. 22. 36.	36. 44.	59
35	Match box stand	Birch	5. 7. 12. 8. 9. 10. 56. 14. 16. 1. 2. 6. 13. 59. 58. 66. 42. 40. 22.	66.	60
36	Round stick	Do.	5. 7. 12. 9. 10. 63. 17. 18. 6. 13. 22.	63.	61
37	Set-square	Do.	5. 7. 12. 8. 9. 10. 56. 38. 59. 14. 31. 32. 64. 22.	64.	62
38	Pen-box	Do.	5. 7. 12. 8. 9. 10. 59. 58. 65. 66. 56. 2. 33. 22.	65.	63
39	Stool	Pine	5. 7. 12. 8. 46. 9. 39. 10. 34. 32. 14. 15. 6. 13. 68. 67. 20. 16. 1. 24. 42. 3. 4. 43. 33. 22.	67. 68.	65
40	Square	Beech	5. 7. 12. 8. 9. 10. 59. 69. 66. 14. 33. 22. 36.	69.	66
41	Drawing-board	Pine	5. 7. 12. 8. 9. 46. 39. 10. 59. 70. 47. 34. 32. 33. 71. 25. 40. 42. 24. 1. 51. 22.	70. 71.	68
42	Marking-gauge	Beech	5. 7. 12. 8. 9. 10. 72. 24. 33. 2. 1. 6. 18. 13. 22. 40.	72.	69
43	Bracket	Birch or Alder	5. 7. 12. 8. 9. 46. 39. 10. 34. 15. 24. 45. 13. 22. 81. 33. 16. 1. 4. 72. 64. 14. 42. 75. 22. 41.	75. 81.	71
44	Picture-frame	Do.	5. 7. 12. 8. 9. 10. 59. 73. 39. 33. 74. 46. 56. 25. 75. 14. 19. 24. 2. 3. 13. 22.	73. 74.	73

	Name of Model.	Wood employed.	Exercises involved.	New Exercises.	Total number of Exercises performed.
45	Stand for tools	Pine	5. 7. 12. 8. 46. 9. 39. 10. 59. 58. 14. 15. 23. 16. 6. 2. 1. 13. 76. 42. 32. 24. 18. 43. 33. 22.	76.	74
46	Tea-tray	Do.	5. 7. 12. 8. 9. 10. 46. 39. 31. 32. 79. 14. 24. 15. 1. 6. 16. 13. 33. 18. 40. 42. 43. 22.	79.	75
47	Book-shelves	Do.	5. 7. 12. 8. 9. 10. 59. 15. 24. 29. 16. 6. 2. 13. 22. 81. 33. 18. 39.	0.	75
48	Hooped bucket	Do.	5. 31. 32. 82. 10. 28. 18. 12. 50. 3. 29. 83. 40. 84. 15. 16. 4. 34. 13. 22.	50. 82. 83. 84.	79
49	Cabinet	Do.	5. 88. 12. 46. 39. 8. 9. 10. 34. 76. 65. 42. 38. 71. 43. 33. 72. 77. 78. 22.	77. 78. 88.	82
50	Small table	Do.	5. 88. 12. 9. 10. 8. 85. 32. 49. 13. 33. 39. 46. 66. 86. 24. 34. 87. 22.	49. 85. 86. 87.	86

Lists of Tools required for different numbers of Pupils.

A. List of Tools required for one pupil.

1 Shooting-board.
2 Handscrews.
1 Metre-measure or Rule.
1 Marking-point.
1 Marking-gauge.
1 Cutting-gauge.
1 pair of Compasses.
1 Square.
1 Bevel.
1 Saw-set.
1 Saw-sharpening clamps.
1 Triangular file.
1 Bow-saw.
1 Dove-tail saw.
1 Turn-saw (broad-webbed).
1 Turn-saw (narrower webbed).
1 Hand-saw.
1 Tenon-saw.
1 Compass-saw.
1 Groove-saw.
1 Axe.
1 Knife.
1 Draw-knife.
½ set (6) Firmer-Chisels.
2 Mortise Chisels.
½ set (6) Gouges.
1 Spoon-iron.
1 Jack-plane.
1 Trying-plane.
1 Smoothing-plane.
1 Compass-plane.
1 Old woman's tooth-plane.
1 Spokeshave.
1 Flat file.
1 Half-round file.
1 Round file.
1 Scraper.
1 Brace, with set of bits.
1 Bradawl.
1 Mallet.
1 Hammer.
1 pair of Pincers.
1 Wire-cutter.
1 pair flat-jawed Pliers.
1 pair round-jawed Pliers.
1 Screwdriver.
1 Glue-pot and Brush.
1 Grindstone.
1 Oilstone.
1 Oil-can.
Sandpaper.

B. Minimum number of Tools required for the simultaneous instruction of 6 to 8 pupils.

1 Shooting-board.
2 Handscrews.
6 to 8 Metre-measures or Rules.
2 Marking-points.
3 to 4 Marking-gauges.*
3 pair Compasses.
3 to 4 Squares.*
1 Bevel.
1 Saw-set.
1 Saw-sharpening clamps.

[* With regard to the minimum number of tools required, the reader is referred to Chapter I., p. 24, and is strongly recommended to provide each child, if possible, with a complete *bench set*, viz :—knife, jack-plane, trying-plane, smoothing-plane, square, marking-gauge, compass, rule or metre-measure, and scraper.—Trs.]

2 Triangular files.
2 Bow-saws.
1 Dove-tail saw.
1 Turn-saw (broad-webbed).
2 Turn-saws (narrower webbed).
1 Hand-saw.
1 Tenon-saw.
1 Compass-saw.
1 Groove-saw.
1 Axe.
6 to 8 Knives.
1 Draw-knife.
1 set (12) Firmer Chisels.
½ set (4) Mortise Chisels.
½ set (6) Gouges.
3 Spoon-irons.
2 Jack-planes.*
3 Trying-planes.*
3 Smoothing-planes.*
1 Compass-plane.
1 Old woman's tooth-plane.
2 Spokeshaves.
1 Flat file.
2 Half-round files.
1 Round file.
3 Scrapers.*
1 Brace, with set of bits.
1 Bradawl.
3 Mallets.
2 Hammers.*
1 pair Pincers.
1 Wire-cutter.
1 pair flat Pliers.
1 pair round Pliers.
2 Screwdrivers.
1 Glue-pot and Brush.
1 Grindstone.
2 Oilstones.
1 Oil-can.
Sandpaper.

C. Minimum number of Tools required for the simultaneous instruction of 12 pupils.

2 Shooting-boards.
3 Handscrews.
12 Metre-measures or Rules.
4 Marking-points.
8 Marking-gauges.*
2 Cutting-gauges.
6 pair of Compasses.
8 Squares.*
2 Bevels.
2 Saw-sets.
2 Saw-sharpening clamps.
4 Triangular files.
3 Bow-saws.
2 Dove-tail saws.
2 Turn-saws (broad-webbed).
2 Turn-saws (narrower webbed).
1 Hand-saw.
1 Tenon-saw.
1 Compass-saw.
1 Groove-saw.
1 Axe.
12 Knives.
1 Draw-knife.
1 set (12) Firmer Chisels.
½ set (4) Mortise Chisels.
½ set (6) Gouges.
3 Spoon-irons.
4 Jack-planes.*
8 Trying-planes.*
8 Smoothing-planes.*
1 Compass-plane.
4 Spokeshaves.
1 Flat file.
3 Half-round files.
2 Round files.
6 Scrapers.*
1 Brace, with set of bits.
2 Bradawls.
4 Mallets.
3 Hammers.*
1 pair of Pincers.
1 Wire-cutter.
1 pair flat Pliers.
1 pair round Pliers.
6 Screwdrivers.
1 Glue-pot with Brush.
1 Grindstone.
2 Oilstones.
1 Oil-can.
Sandpaper.

* See note, page 201.

D. **Complete list of Tools required for the simultaneous instruction of 12 pupils.**

3 Shooting-boards.
4 Handscrews.
12 Metre-measures.
8 Marking-points.
12 Marking-gauges.
2 Cutting-gauges.
6 pair of Compasses.
12 Squares.
2 Bevels.
2 Saw-sets.
2 Saw-sharpening clamps.
4 Triangular files.
4 Bow-saws.
3 Dove-tail saws.
4 Turn-saws (broad-webbed).
4 Turn-saws (narrower webbed).
3 Hand-saws (tenon-saws).
1 Groove-saw.
2 Axes.
12 Knives.
2 Draw-knives.
2 sets (24) Firmer Chisels.
1 set (8) Mortise Chisels.
1 set (12) Gouges.
6 Spoon-irons.
12 Jack-planes.
12 Trying-planes.
12 Smoothing-planes.
1 Compass-plane.
1 Old woman's tooth-plane.
6 Spokeshaves.
4 Flat files.
6 Half-round files.
4 Round files.
6 Scrapers.
1 Brace with set of bits.
3 Bradawls.
8 Mallets.
12 Hammers.
2 pair of Pincers.
1 Wire-cutter.
1 pair flat Pliers.
1 pair round Pliers.
6 Screwdrivers.
1 Glue-pot with Brush.
1 Grindstone.
3 Oilstones.
1 Oil-can.
Sandpaper.

The cost of providing the above tools, calculated according to the prices* now current (in Sweden), is as follows :—

List	(*A*)	about	50	Kronor.†
,,	(*B*)	,,	85	,,
,,	(*C*)	,,	135	,,
,,	(*D*)	,,	205	,,

The number of benches required is as follows :—

For	List	(*A*)	1
,,	,,	(*B*)	3 or 4
,,	,,	(*C*)	8
,,	,,	(*D*)	12

If double benches are used, only half the number will be required in the cases of (*B*), (*C*), and (*D*) respectively. As double benches are cheaper in proportion to single benches, they may in some cases be preferred to single ones.

Tools of the best quality should always be procured; they are the cheapest in the long-run.

It is also desirable that drawings and constructions of the models should be procured.

In connection with this, it may be mentioned that the annual cost of timber and other materials in a small country school in Sweden, where 10 to 15 children receive instruction, is from 10 to 20 kronor, exclusive of the outlay of replacing worn-out tools.

The same materials for a class of 16 boys in England would cost about £1 10s. annually. Red deal pine is the best soft-wood for Slöjd, and can be obtained at any good timber yard. Lime-tree, sycamore, and chestnut all make good substitutes for birch. The two former are very fine in the grain, and are good for scoops, bowls, etc. They are a little dearer than birch. American canary wood can also be recommended as a wood suitable for flat articles ; it is a little harder than deal, and works easily. This wood can be obtained anywhere, and is not dear.

---

* The English prices for the tools are given on page 214.

† The Swedish krona is worth 1s. 1½d. Eighteen kronor = £1.

INDEX.

H

I

J

K

L

M

N

O

P

R

S

Spotted something that isn't quite right?

Let us know and we'll fix it for future reprints.
And as a thank you for your diligence, we'll
send you a free Sothis book of your choice.

publisher@sothis.press

This page intentionally left blank

Except of course for that line above. And this one.
And whilst we have your attention, perhaps you'd
like to follow us on Instagram @sothis.press. Go on.

Made in the USA
Coppell, TX
05 November 2026

11365027R00134